FIRST MEN O

On July 16, 1969, Apollo 11 was launched into outer space. Its destination: the moon.

Four days later, the spiderlike Lunar Module touched down on the moon's forbidding surface. It rested there for almost seven hours before Neil A. Armstrong made man's first footprint on the moon. He was followed some 20 minutes later by Edwin E. Aldrin, Jr.

On July 24, the space capsule became a fiery ball for a few breathtaking moments as it hurtled into the earth's atmosphere. Then the parachutes opened and Apollo 11 splashed down in the Pacific.

HERE, IN WORDS AND STUNNING PICTURES, IS THE STORY OF MAN'S BOLDEST ACHIEVEMENT—BRILLIANTLY CAPTURED FOR NOW AND FOR FUTURE GENERATIONS.

PUBLISHER'S NOTE

Here, in response to the worldwide interest in what is unquestionably the most spectacular scientific achievement in history, Bantam Books, in conjunction with *The New York Times*, is making **WE REACH THE MOON** available to readers across the nation and throughout the world less than 72 hours after the splashdown of Apollo 11.

WE REACH THE MOON, the result of nearly two years of research and writing, is the *first* complete report of the American moon efforts and accomplishments—an original work not only of great scientific interest but of genuine literary merit and broad social significance as well. This edition, with its dramatic color photographs, has been printed and distributed through the techniques pioneered by Bantam Books.

ABOUT THE AUTHOR

JOHN NOBLE WILFORD is the leading aerospace reporter for *The New York Times*. He has covered every phase of the space program and every Project Apollo shot leading up to the epochal moon landing. **WE REACH THE MOON** is Mr. Wilford's definitive account of the incredible space achievement, from its beginnings with the faint beep-beep of Sputnik to its conclusion at the Apollo 11 splashdown.

WE REACH THE MOON

The New York Times
STORY OF MAN'S GREATEST ADVENTURE
By John Noble Wilford

Illustrated with line drawings and photographs

BANTAM BOOKS
NEW YORK · TORONTO · LONDON

WE REACH THE MOON

A Bantam Book / published July 1969
2nd printing
3rd printing

ACKNOWLEDGMENTS

Grateful acknowledgment is made for permission to reprint the
following copyrighted material:
Narrative by John W. Finney beginning on page 47: © *1962 by*
The New York Times Company. Reprinted by permission.
Narrative by John N. Wilford beginning on page 99: © *1967 by*
The New York Times Company. Reprinted by permission.
Narrative by Harry Schwartz on page 108: © *1967 by The New*
York Times Company. Reprinted by permission.
Narrative by Christopher Kraft beginning on page 180: © *1969 by*
The New York Times Company. Reprinted by permission.
Prose poem by Archibald MacLeish beginning on page 205: ©
1968 by The New York Times Company. Reprinted by permission.
Narrative by Tom Buckley beginning on page 208: © *1969 by The*
New York Times Company. Reprinted by permission.
Narrative by Bernard Weinraub beginning on page 258: © *1969*
by The New York Times Company. Reprinted by permission.

Published simultaneously in the United States and Canada

Bantam Books are published by Bantam Books, Inc., a National
General company. Its trade-mark, consisting of the words "Bantam
Books" and the portrayal of a bantam, is registered in the United
States Patent Office and in other countries. Marca Registrada.
Bantam Books, Inc., 271 Madison Avenue, New York, N.Y. 10016.

PRINTED IN THE UNITED STATES OF AMERICA

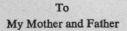
To
My Mother and Father

Contents

Acknowledgments

This is, first and foremost, a newspaper reporter's book. I am not a scientist or engineer. But my years of covering the Apollo Project were an enlightening adult-education course in the fields of knowledge I either neglected as a university student or was not even exposed to in the pre-space curriculum. Nor am I a certified historian. But no journalist who looks beyond his typewriter can ignore the elements of history in the raw material that is his story.

And I am not—I was about to say, I am not a romantic, a dreamer, for journalists pride themselves in being detached observers of other men's passions. But who could cover the story of man's first footsteps on another world, and cover it as more than a science story or another nuts-and-bolts technology story, without feeling in himself a flutter of all the romantic urges that have sent men across oceans, up mountains and out into the air and then the space beyond the air? I could not.

Whatever future historians may say of Apollo, I will be grateful to *The New York Times* for the opportunity to be a close observer of this great adventure.

This is not solely my book. I owe a large debt to a number of members of the *Times* staff, including Walter Sullivan, Henry R. Lieberman, Evert Clark (no longer with the paper), Tom Buckley, Harry Schwartz, John W. Finney, Richard Witkin, Harold M. Schmeck, Jr., Sandra Blakeslee, Bernard Weinraub, James T. Wooten and Douglas Kneeland. Their stories and editorial assistance were an invaluable contribution to the book. The diagrams are the work of the *Times* map department headed by Andrew Sabbatini.

I want to single out for credit Richard D. Lyons, a boon companion and especially helpful colleague through all the major Apollo missions, and William K. Stevens, who provided most of the material and writing for the chapter on the Apollo 11 crew. Another good companion and colleague through all this was Douglas M. Dederer, who was especially helpful during my many visits to Cape Kennedy. The National Aeronautics and Space Administration and many of the major Apollo industrial contractors helped make available the many documents and background materials that facilitated coverage of Apollo.

Others who made the book possible were Carter Horsley and Sydnor Vanderschmidt, who made important research contributions. Mark Bloom was kind enough to read and offer suggestions on a part of the book. Mary Rourke typed much of the manuscript.

No one gave me more encouragement than my wife, Nancy, who shares my enthusiasm for the story and was uncomplaining of the time taken from our lives by the program to get men to the moon.

Every writer needs an editor, and I was fortunate to have two very skillful and understanding editors—Lee Foster and Donald Johnston. Finally, I want to thank Jack Stewart, who had the original idea, for his encouragement and patience over the long months of giving birth to this book.

JNW
July 1969

Earth is the cradle of the mind, but one cannot live in a cradle forever.

Konstantin E. Tsiolkovski
Russian rocket pioneer

I believe that this nation should commit itself to achieving the goal, before the decade is out, of landing a man on the moon and returning him safely to earth.

John F. Kennedy
Special Message to Congress
May 25, 1961

*Have we not lately in the moon
Found a new world, to the old unknown?
Discover'd seas and lands Columbus
And Magellan could never compass?*

Samuel Butler
English poet

Principal Dramatis Personae

In Order of Their Appearance

Neil A. Armstrong, Edwin E. Aldrin, Jr., Michael Collins	The astronauts who did it.
Richard M. Nixon	He made the historic phone call.
Wernher von Braun	Director of the Marshall Space Flight Center at Huntsville, star of the German rocket specialists, and one of the most charismatic of the space pioneers.
John F. Kennedy	He made the commitment in 1961 to put an American on the moon "before this decade is out."
Lyndon B. Johnson	As Vice President and head of the Space Council, he encouraged the Kennedy decision; later, as President, he pushed the space effort.
James E. Webb	NASA's no. 1 man until 1968, the administrator who made the vast program work despite endless challenges.

Robert C. Seamans, Jr.

NASA's associate administrator, or general manager for mobilization of the immense Apollo team; he later became deputy administrator.

Alan B. Shepard, Jr.

His success as the first man in space—with a suborbital flight —contributed to the Kennedy decision to go "all out."

Hugh L. Dryden

NASA's deputy administrator, or no. 2 man, during the formative years, and chief liaison with the scientific community.

William H. Pickering

As director of the Jet Propulsion Laboratory, he directed much of the unmanned-satellite work.

Brainerd Holmes

As NASA's first administrator for manned flight, he directed Apollo in the early years, but later resigned because of a dispute over funding.

John C. Houbolt

An obscure NASA engineer whose prodding led to the adoption of the lunar-orbit rendezvous method for a moon landing.

Joseph Shea

Apollo spacecraft manager, an easygoing punster whose systems and equipment analyses laid the engineering foundations for Apollo.

Robert R. Gilruth

Director of the Manned Spacecraft Center at Houston, with overall responsibility for development of the spacecraft, flight crew training and flight control.

Christopher C. Kraft — Director of flight operations at Houston, he ran the missions and was ready for any emergency.

Maxime Faget — Principal designer of the Mercury capsule.

Kurt H. Debus — Director of the Kennedy Space Center and one of the von Braun Germans, he laid out the moonport and was responsible for launchings.

George E. Mueller — He replaced Holmes as administrator for manned flight in 1963.

Samuel C. Phillips — A tough-minded air force general who, after being named Apollo program director in 1964, emerged as the project's driving force and key decision-maker.

John H. Glenn, Jr. — The first American to make an orbital flight, he caught the national imagination and became the prototype of the new Space Age folk-hero.

Dr. Charles A. Berry — Flight surgeon, whose word on the physical status of the astronauts was law before and during missions.

Virgil I. Grissom, Edward H. White, Roger B. Chaffee — The astronauts who died in the 1967 launching-pad fire that set back the Apollo program nearly two years.

Rocco A. Petrone — A former football player who directed the mammoth launch operations at Cape Kennedy.

Arthur Rudolph	Another German rocketeer, he was largely responsible for overcoming the countless problems in building Saturn 5.
Joseph G. Gavin, Jr.	A soft-spoken Grumman engineer who directed development of the Lunar Module.
Dale D. Myers	The calm but tough engineer who guided the North American Rockwell team through painstaking job of building the Command Module.
George M. Low	A NASA veteran who drew the difficult job after the fire of rebuilding a safer ship for the moon landing.
Charles Stark Draper	The nation's top authority on inertial guidance. As director of M.I.T.'s Instrumentation Laboratory, he supervised development of the vital guidance and navigation system.
Thomas O. Paine	A sharp and highly respected advocate of the lunar-landing project, he succeeded Webb as NASA boss.

. . . AND . . .

Other astronauts and their families, scientists and technicians at the various space centers, mission directors and mechanics, flight trackers and communications men, executives and employees of the thousands of contractors and subcontractors, Congressmen and government officials, and the American people.

Prologue

Footprints On The Moon

On the desolate, lifeless landscape of the moon, a funny-looking vehicle squats motionless under the sun's glaring rays. On one of its four spindly legs is attached a small, stainless steel plaque which reads:

HERE MEN FROM THE PLANET EARTH
FIRST SET FOOT UPON THE MOON.
JULY 1969 A.D.
WE CAME IN PEACE FOR ALL MANKIND.

The vehicle, the dispensable lower half of a lunar landing craft, is a monument to the historic event on July 20 when two American astronauts planted the first human footsteps on the earth's only satellite. The monument will remain there for ages, for without air the moon has no corrosion, without water it has no erosion.

The man who took the first step was Neil A. Armstrong, the 38-year-old civilian commander of the Apollo 11. As he reached the bottom of the landing craft's ladder and extended his booted left foot to touch the moon's powdery surface, he capsulized the momentous meaning of the moment: "That's one small step for a man, one giant leap for mankind."

He was followed down the ladder minutes later by Edwin E. Aldrin, Jr., a 39-year-old Air Force colonel. For two hours and 21 minutes, the two men, cautiously at first and then boldly, capered about on the barren, rock-strewn lunar terrain. They tested their agility in the alien environment and took photographs. They erected scientific experiments and collected rock and soil samples. They set up a television camera so the whole world could watch. At one point Armstrong said to his colleague, "Isn't this fun?"

All the while the third member of the crew, Michael

Collins, 38, an Air Force lieutenant colonel, piloted the Command Ship in lunar orbit 70 miles above the surface, waiting for the two explorers to rejoin him for the trip back to earth. Altogether, the lunar visit lasted 21 hours and 37 minutes.

An Incredible Triumph

For the Apollo 11 crew, the successful 500,000-mile mission involving 88 separate steps was an incredible triumph of skill and courage. For the United States space team, from the other astronauts down to the technicians, it was the fulfillment of a decade of technological striving. For the world, it was the most dramatic demonstration of what man can do if he applies his mind and resources with single-minded determination. The moon, long the symbol of the impossible, was now within man's reach, the first port of call in the new age of spacefaring.

There have been other daring expeditions, of course. But Apollo 11 was different: a large proportion of mankind bore it witness. Through television and radio, literally hundreds of millions of people followed the activities aboard Columbia, the Command Ship, and Eagle, the landing craft —names chosen because, as Armstrong put it, they were "representative of the flight and the nation's hope." The television transmissions across the 238,000 miles from the moon were so clear and sharp, accentuating the deep shadows and bright sunlight, that they often created a sense of unreality.

Nor was Apollo 11 really a voyage of geographical discovery (its destination was known and had been photographed and probed by unmanned satellites) or an expedition of conquest (the 1967 space treaty, which was signed by the United States, stipulates that outer space, including the moon, "is not subject to national appropriation"). The space agency's primary objective was simply: "Perform a manned lunar landing and return."

Though the mission was accomplished almost without flaw, it was never without suspense and anxiety. The astronauts faced risks on the moon never before encountered by man. And, as with all space flights, chances of failure and disaster were ever present—the blast-off of the giant Saturn 5 rocket at Cape Kennedy, the entry of the spaceship into earth and lunar orbits, the never-before-attempted landing and lift-off from the moon, the link-up of Columbia and Eagle, the re-entry into the atmosphere, the splashdown. An error or malfunction of any of the millions of individual parts anywhere along the way could have ended the mission short of the goal. An equipment failure or accident on the moon could have left the astronauts stranded.

Heroes

But they made it. After eight days in space, they splashed down in the Pacific to a presidential greeting aboard the recovery carrier, the U.S.S. *Hornet*. They were the heroes of the nation and the world.

"This is the greatest week in the history of the world since the Creation," a jubilant President Nixon told the space travelers. "As a result of what you've done, the world has never been closer together before."

The welcome was not the familiar red-carpet ceremony of space missions in which the returning astronauts walk across the deck of the carrier. The Apollo 11 crewmen were inside an isolation van and they conversed with Mr. Nixon by microphone. The astronauts were under quarantine as a precaution against the possibility that they brought back some deadly lunar organisms for which man has no immunity.

People everywhere, the important and the common, hailed the achievement. Perhaps never before had one event so captured the world's imagination and spirit of adventure. There were some grumbles about problems to be solved here on earth, but for days Apollo monopolized the headlines, TV sets and radios, even in countries with Communist governments ordinarily hostile to the United States. The Soviet Union reported the mission at unprecedented length. "We rejoice at the success of the American astronauts," the government newspaper *Izvestia* said. Both Africans and Laplanders followed the mission on transistor radios.

Scores of heads of state and government sent President Nixon congratulatory messages praising the Apollo 11 crew for having added a new dimension to man's knowledge. Many shared the sentiment of Mrs. Golda Meir, Israel's premier, who hoped the demonstration of man's capacity might "open the way to that era of universal peace presaged by the prophets of old."

Culmination of Long Effort

Apollo 11 was the culmination of 12 years of effort since man discovered he could fly in space, and eight years of the Apollo program since President John F. Kennedy, concerned about Soviet space successes, committed the nation in 1961 to landing a man on the moon "before this decade is out." That commitment set in motion the greatest mobilization of men and resources ever undertaken for a peaceful project of science and exploration. The building of rockets and spacecraft required the concerted efforts of 20,000 industrial contractors, scores of university laboratories, and 400,000 people. In terms of money, the program cost $24 billion for Apollo, not counting the $392 million for the Mercury flights and the $1.3 billion for the Geminis—the two forerunner projects in which men proved they could live and work in space.

As the program gained momentum, it brought great changes to many rural regions and small towns where mills and abandoned war plants were converted to making such material as guidance instruments and booster rockets. Most of the activity was concentrated in the so-called "fertile crescent" stretching from Florida to Texas. At one end, engineers erected a billion-dollar moonport on the sand around Cape Canaveral; at the other, bulldozers cleared ranchlands outside Houston for the complex flight control center, the heart of the system. And across the globe, at remote spots and on ships, was strung a network of radar stations to track the flights.

Though the program was carefully conceived and elaborately planned, mistakes were made and accidents happened. The three astronauts who were to make the first test flight of the Apollo Command Ship—Virgil I. Grissom, Edward H. White and Roger B. Chaffee—died in a fire that broke out in their cockpit during a launching-pad rehearsal on January 27, 1967. The tragedy plunged the space team into its gloomiest period, a time of self-doubt from which it took nearly two years to recover.

Man's going to the moon sprang from political and technological considerations and even deeper, from the very roots of the human spirit. One consideration was national security. Space was clearly connected with development of rockets and instruments vital to defense. Another was national prestige: the challenge to a great power to lead in an age of technology. What moved the human spirit was, as Aldrin put it in a reflective moment on the way home, "the insatiable curiosity of all mankind to explore the unknown" that had sent explorers to brave the harshest conditions of this planet—to the poles, to the tops of mountains, to the loftiest balloon altitudes, to the deepest portions of the sea. Then why not the moon?

There were critics who said that the Apollo program cost too much, that the money and talent could be more usefully directed to fighting disease and poverty, that it was a "childish stunt" to make a race out of going to the moon and insisting on an artificial deadline. But by the time of the lunar mission, thought turned to prospects that the astronauts might bring back some clues to help solve the mysteries of the universe.

Whatever its impact on man and earth, the voyage of Apollo 11 marked a bold new extension of man's dominion over his environment. It broke the bonds of his native planet and enabled him to reach another world.

Chapter I

Challenge and Commitment

The long countdown for the voyage to the moon began in the spring of 1961, in the fourth year of the Space Age.

Americans were still haunted and not a little humiliated by the fact that it was the Russians, not they, who had opened up the vastness of space to direct exploration by man. Americans could not forget the dark headlines on Saturday morning, October 5, 1957, announcing that the Soviet Union had launched the first artificial earth satellite. The satellite weighed 183.6 pounds and was 22.8 inches in diameter, an aluminum sphere with antennas like cat's whiskers that broadcast faint beep-beep-beeps heard round the world. This was Sputnik 1. For scientists, Sputnik meant one thing. It was a new instrument in man's quest for knowledge of his universe, the first of many far-ranging vehicles of discovery. For the laymen, from presidents to premiers on down, the beeps carried another message, startling but clear: Made in the Soviet Union.

Overnight the world's image of the Soviet Union changed. The Soviet Union had emerged from World War II victorious but badly battered. The Soviet Union was the capital of world communism but unable to fill adequately the everyday needs of its working class. Thus the stereotype was one of a great but still backward giant. But the giant, it became clear, was not as backward as the West had assumed. In a report to Washington, the United States Information Agency said that Sputnik represented "a major watershed in the Western European evaluation of the relative power standing of the United States and the Soviet Union."

Americans were suddenly jarred out of what the British writer C. P. Snow called their "technological conceit," which was based in part on the belief that science and

1

totalitarianism were incompatible. The scoffing ceased. No longer did Americans make condescending cracks about the poor quality of Russian ball-point pens or the exaggerated Soviet claims about inventing nearly everything from radio to the proverbial kitchen sink. Those who had been skeptical now believed the Russians' contention in August, 1957, that they had developed an intercontinental ballistic missile capable of pounding American cities. If they could launch Sputnik, they could fire an accurate ICBM. Consequently, though some officials in Washington disparaged Sputnik as a "hunk of iron" or a "silly bauble," Americans generally understood its challenge.

Sputnik signaled that the Soviet Union had arrived as a world power with a capability second to none in science and technology. And Sputnik, more than any other single postwar event, sharpened the world's awareness of science and technology as symbols as well as sources of strength. Nikita S. Khrushchev, the Soviet premier, drove the implications home in an interview on October 7 with James Reston of *The New York Times*. The Soviet Union, Khrushchev said, had "every kind of rocket required for modern war." A new dimension had been added to power politics, and a new arena—space.

The American response reflected the seriousness of the challenge. In November, President Dwight D. Eisenhower, who at first had dismissed Sputnik as an event of "scientific interest," said sweeping changes were being made in the Defense Department to give top priority to missile and rocket projects. At the same time the Defense Department ordered the army to rush to the launching pad a Jupiter-C rocket and Explorer satellite whose development had until then been discouraged. The word in Washington was: Sure we're behind, but it's only temporary.

By this time, however, the Russians had already launched, on November 3, 1957, an even more impressive Sputnik, weighing 1,120 pounds and carrying a live dog, a sure harbinger of Soviet efforts to put men into space. Moreover, the first American attempt to break out of the atmosphere ended in failure. Vanguard blasted off on December 6, rose a few feet and then collapsed in smoke and flame. Not until January 31, 1958, did the United States orbit a satellite. It was Explorer 1, only 18.13 pounds but the discoverer of the Van Allen radiation belts around the earth.

In the next three years, through the spring of 1961, the Soviet Union let the world know that Sputnik 1 and Sputnik 2 were not isolated stunts. Subsequent Soviet spacecraft achieved breakthroughs by hitting the lunar surface and photographing the far side of the moon. In 1960 Sputnik 5 carried two dogs into orbit, both of which were recovered in the first demonstration of man's ability to bring living creatures safely back to earth. Even in their attempts to fly by the planets Venus and Mars, which were failures, the Russians proved their deep commitment to space exploration.

Meanwhile, the American program was gathering force. Vanguard got into orbit on March 17, 1958. Both houses of Congress created space committees. On July 29, 1958, President Eisenhower signed the National Aeronautics and Space Act establishing a civilian space agency, which went into operation on October 1. There were more Explorers, a series of far-ranging craft called Pioneers, experimental satellites for communications transmission and navigation aids. Project Mercury, the nation's first manned spaceflight program, was announced in the fall of 1958, the first seven astronauts chosen and the unmanned test launchings begun.

Time of Introspection

Even so, the United States in the spring of 1961 lagged behind the Soviet Union in space exploration. Could it ever catch up? This was only one of the questions Americans were asking themselves and all were related to the basic question: What is wrong with America? For Americans, prompted in no small way by the Sputnik humiliation, were beset by self-doubts, a new and unusual mood of introspection.

They were beginning to have serious second thoughts about the aura of self-satisfaction described in 1959 by political columnist Walter Lippmann:

> The critical weakness of our society is that for the time being our people do not have a great purpose which they are united in wanting to achieve. The public mood

of the country is defensive, to hold on to and to conserve, not to push forward and to create. We talk about ourselves these days as if we were the completed society, one which has achieved its purposes, and has no further business to transact.

By 1961 Americans were waking up to the unfinished business. Speaking of this post-Sputnik time, Wernher von Braun, the rocket expert, later summarized: "It became popular to question the bulwarks of our society, our public education system, our industrial strength, international policy, defense strategy and forces, the capability of our science and technology."

Americans devoted their public forums and cocktail hours to the criticisms leveled in John Kenneth Galbraith's popular book *The Affluent Society*. The Harvard economist condemned the way Americans, enjoying an affluence unmatched in history, were accentuating personal comfort and pleasure at the expense of public needs—the poor, the sick, the festering slums, crowded schools and creaking urban transportation systems. This theme of flabbiness also ran through a series of articles entitled "The National Purpose," which were published jointly in 1960 by *Life* and *The New York Times*. The articles reflected the fact that Americans were worried about themselves. They worried as never before that their schools were inadequate, that their children might be watching too much television when they should be studying more science, that their pursuit of the good life might be shortsighted, that the nation's black people and poor people were growing explosively restless from years of neglect.

It was also disquieting for Americans when they looked abroad. Sputnik had dealt their self-esteem and self-confidence one obvious blow. But the cold war had also taken a subtle toll. Conflict seemed to be epidemic and largely beyond U.S. control. There appeared to be no formula for settlement in Berlin or Vietnam, the Middle East or Cuba, Kashmir or Laos, that did not give birth to a new crisis or, at best, a dubious truce. The old formulas led only to frustration. The new ones, if perceived at all, required a drastic remolding of popular concepts of international challenge and response. In their attempts to understand the new situation, politicians, diplomats and scholars were wont to quote Henry A. Kissinger, a Harvard professor who would later become an influential White House adviser. In his book *Nuclear Weapons and Foreign Policy*, Kissinger said that survival for America "depends not only on our strength, but also on our ability to recognize (and fight) aggression in all its forms." If so, were not the Russian space achievements an implied technological threat? Most people believed they were and reacted accordingly. Insofar as there was dread and alarm it no doubt stemmed from a fear of surprise born of misunderstanding the new shape of international conflict. As Kissinger wrote, "In the nuclear age, by the time a threat has become unambiguous it may be too late to resist it."

All this was a new and confusing experience for Americans, who had a reputation as cocky optimists and had, as British historian D. W. Brogan once observed, grown up with "the illusion of American omnipotence."

"Let the Word Go Forth"

Out of such a national mood—compounded of self-doubt, vague and almost puritanical longings to purge past self-indulgence with hard work, and a restlessness to do greater things—came the decision by Americans to reach for the moon.

John Fitzgerald Kennedy had just become president and in the spring of 1961 was groping for ways to fulfill his famous campaign pledge to "get the country moving again." He was 43 years old, the first president born in this century, and imbued with the eagerness and energy of youth. "Let the word go forth," he declared in his inaugural address, "that the torch has been passed to a new generation of Americans, born in this century, tempered by war, disciplined by a hard and bitter peace. . . ." He went on to make an eloquent appeal to Americans to accept hard tasks and live up to their potential for greatness.

This made his first four months in office all the more disappointing to him. His legislative program languished in the hoppers of Congress. The economy remained sluggish. And in less than one week in April both his administration and the nation were jarred by another Soviet space exploit and humiliated only 90 miles offshore, in Cuba. On April 12 the Soviet Union jumped further ahead in space by launching the first man into orbit, Yuri A. Gagarin. Then, on April 17, came the Bay of Pigs invasion. Not only did this American-backed assault on Cuba fail to oust Premier Fidel Castro but, through misdirection, it turned into an utter rout. All this, it was reported, led the shaken new president to tell Republican Senator Barry Goldwater: "What a lousy, fouled-up job this has turned out to be."

It would be an oversimplification to conclude that Kennedy decided on the man-to-the-moon program solely as a direct response to that disastrous April. It was no doubt an important factor. But it was no spur-of-the-moment decision. The groundwork had been laid much earlier. In his election campaign Kennedy had declared that space "is our great new frontier." And, of course, "new frontier" had become the catchphrase of the Kennedy administration. Although he gave no hint in his inaugural speech of a major new space program, the National Aeronautics and Space Administration already had under consideration three alternative long-range plans for activities beyond Project Mercury. They included: (1) further scientific exploration of the moon and the planets with unmanned vehicles; (2) emphasis on practical applications of space through communications and weather satellites; (3) concentration on a manned landing on the moon.

Science advisers in the Eisenhower administration had viewed a manned lunar landing as a "scientific luxury." Nor were they encouraged to think otherwise. At a cabinet meeting in the fall of 1960, George B. Kistiakowsky, Eisenhower's science adviser, described a study of manned lunar landings that he had conducted. The cost, he estimated, would be about $35 billion. There was a gasp of disbelief around the cabinet table. Would anyone seriously consider such an undertaking? Even the man who was to be Kennedy's science adviser, Jerome B. Wiesner of the Massachusetts Institute of Technology, was not favorably disposed in the beginning. In a report Wiesner made in January to Kennedy, he suggested that the manned space flight was being given too high a priority. It was a feeling shared by many scientists, who feared such a project would drain money from their own research work. But Congress, perhaps reflecting the public mood, was for an expanded manned space program. In July, 1960, the House space committee recommended "a manned expedition to the moon before the end of the present decade."

At about the same time in 1960 NASA called major aerospace contractors to a meeting in Washington at which the rough outlines of a manned program to follow Project Mercury were described. NASA engineers were thinking primarily of missions in earth orbit and around the moon. A lunar landing was not seriously discussed, but a plan was on the drawing boards in case it was needed.

It was not to find strong advocates in high places until Kennedy took office. In one of his first moves Kennedy assigned his vice-president, Lyndon Baines Johnson, to head the National Aeronautics and Space Council, a White House advisory body seldom consulted during the Eisenhower years. Johnson had been studying the possibilities of a manned lunar landing ever since he became interested in space at the helm of a Senate investigation after the Sputnik crisis. To replace T. Keith Glennan as head of NASA, Kennedy appointed James E. Webb, a North Carolina lawyer who had learned his way around Washington as Bureau of the Budget director under Harry S Truman.

Webb immediately ordered a complete reexamination of NASA's long-range plans and discovered that most scientists and engineers in the agency favored an accelerated manned space program. If there were any doubts they all but vanished on April 12, 1961, when Yuri Gagarin orbited the earth. The Russian spectacular brought new pressure on Webb, Johnson and Kennedy to show that the United States did not intend to forfeit the space race to the Soviet Union. Congress called in Webb and demanded to know if America could possibly land men on the moon by 1967. According to Robert C. Seamans, Jr., who was NASA's associate administrator at the time, the House Committee on Science and Astronautics "held a highly emotional hearing" the day after Gagarin's flight. The congressmen pleaded for more space efforts, while Webb was forced to defend the administration's policies, however ill-defined they then were. Two days after the Soviet flight, George M. Low, another NASA official, was interrupted during his testimony before the same House committee by Representative David S. King of Utah. As Seamans recalls it,

The Congressman quoted a story from the Bible in which an army of 10,000 marched into battle against a superior force of 50,000. He asked if a similar situation existed in space, and whether the United States would be decimated in similar fashion. Specifically, he asked whether the Soviets were planning a manned lunar landing in 1967 in order to commemorate the 50th anniversary of the Russian Revolution. We pleaded ignorance of such plans. It was then asked if the United States could land men on the moon in 1967, and, if so, at what cost. We stated our belief that such a mission was a possibility in 1967, provided the nation wanted to make such a major commitment. We estimated the cost at between $20 billion and $40 billion.

The seeds were sown. Kennedy called in Johnson on April 20 and asked him to investigate what could be done to catch up with the Russians. In a five-paragraph memorandum Kennedy asked: (1) Where do we stand in space? (2) What are we doing about going to the moon? (3) Is there any other program that promises dramatic results, in which the United States could win? (4) What about American rocket capabilities? (5) Is the American program making a maximum effort?

There was a strong catch-the-Russians mood in Washington. After a series of meetings with officials of the space agency, the Defense Department, the Budget Bureau and private industry, Johnson concluded that a vigorous space program aimed at a manned lunar landing was both desirable and technologically feasible. This was the gist of his report to Kennedy.

On May 5 Commander Alan B. Shepard, Jr., made a short suborbital flight in a Mercury capsule and became the first American astronaut to enter space. As the nation cheered Shepard's safe return, Webb went into a long weekend meeting with Defense Secretary Robert S. McNamara at his Pentagon office. There, along with Pentagon and NASA advisers,* they worked out the final version of the moon-landing plan. On Monday, May 8, they sub-

---

* Seamans and Abraham Hyatt, director of NASA's planning and evaluation staff, were at Webb's side. Roswell L. Gilpatric, deputy secretary of defense, and John H. Rubel, assistant secretary of defense, were McNamara's associates at the meeting. Seamans and Rubel actually drafted the memorandum proposing the lunar-landing goal.

mitted the plan to Johnson, who rushed it to the White House.

"Running through these discussions," Edward C. Welsh, executive secretary of the National Aeronautics and Space Council, recalled later, "was the theme, could we get to the moon, should we if we could, how much would it cost, what else did we need to do if we decided to?"

And NASA's deputy administrator, Hugh L. Dryden, said: "We looked at all the alternatives and we decided Apollo [the moon-landing concept] was the earliest project in which we had a chance of coming out ahead." In a lecture at the Massachusetts Institute of Technology eight years later, in 1969, Seamans summed up the reasoning that had prevailed at that weekend meeting in May. He said:

> It was felt that we had profited from the openness of Alan Shepard's flight, and that it would be a mistake to change the United States policy in favor of a classified manned space effort. Consequently, any goals selected would be immediately known to the Soviets, thus permitting them to redirect their effort in order to surpass us if they wished. If we were to have a reasonable chance of achieving our objectives ahead of the Soviets, the objectives had to require a capability well beyond that of their already-demonstrated launch vehicle and spacecraft. . . . It was felt that the Soviets might uprate their launch vehicle sufficiently for circumlunar flight, and the conclusion was reached that the lunar landing and return was the most appropriate target, since it clearly would require a completely new and greater scale of operation. The question was raised by the Department of Defense as to whether even this bold objective was sufficient; however, NASA officials argued strenuously that we were prepared to undertake the manned lunar landing and return, but were not prepared for manned planetary missions.

Shoot for the Moon

Finally, on May 25, President Kennedy made the trip across Washington from the White House to Capitol Hill to deliver a special message to a joint session of Congress. He described his 47-minute speech as a second state of the union address, thus departing from the custom that such addresses are usually delivered by the president only once a year, in January. "These are extraordinary times," Kennedy began. "And we face an extraordinary challenge. Our strength as well as our convictions have imposed upon this nation the role of leader in freedom's cause."

It was an embattled cause, as anyone could tell by scanning the headlines of *The New York Times* that morning:

27 BI-RACIAL BUS RIDERS JAILED IN JACKSON, MISS.
U.S. MOVES FOR INJUNCTION AGAINST POLICE IN ALABAMA
HOUSE VOTES RISE IN ARMS SPENDING
RED CHINA SAYS U.S. AIM IS TO MAKE LAOS A COLONY
ALGERIA PARLEY STILL SPARRING

People were also talking that day about the deadlocked nuclear test ban treaty negotiations in Geneva, the president's impending trip to Vienna for his first meeting with Khrushchev, the recent military coup d'état in South Korea, the roiling unrest in the Congo and the trial in Jerusalem of Adolf Eichmann, a haunting reminder of man's past inhumanity to man.

Kennedy reserved his most ringing phrases for the main purpose of his speech, his call for a greatly expanded program of space exploration. With the following words he rallied Americans to shoot for the moon in the 1960s:

Now it is time to take longer strides—time for a great new enterprise—time for this nation to take a clearly leading role in space achievements which, in many ways, may hold the key to our future on earth.

I believe we possess all the resources and talents necessary. But the facts of the matter are that we have never made the national decision or marshaled the national resources for such leadership. We have never specified long-range goals on an urgent time schedule; or managed our resources and our time so as to insure their fulfillment.

Recognizing the head start obtained by the Soviets with their large rocket engines, which gave them many months of lead time, and recognizing the likelihood that they will exploit this lead for some time to come, in still more impressive successes, we nevertheless are required to make new efforts of our own.

For while we cannot guarantee that we shall one day be first, we can guarantee that any failure to make this effort will make us last.

We take the additional risk of making it in full view of the world. But as shown by the feat of Astronaut Shepard, this very risk enhances our stature when we are successful.

But this is not merely a race. Space is open to us now. And our eagerness to share its meaning is not governed by the efforts of others. We got into space because whatever mankind must undertake, free men must fully share.

I therefore ask the Congress, above and beyond the increases I have earlier requested for space activity, to provide the funds which are needed to meet the following national goals:

First, I believe that this nation should commit itself to achieving the goal, before this decade is out, of landing a man on the moon and returning him safely to earth. No single space project in this period will be more impressive to mankind or more important for the long-range exploration of space. And none will be so difficult or expensive to accomplish.

This was the heart of the moon commitment. Here were phrases that would be repeated time and again during the decade as fervent incantations explaining, defending and inspiring: key to our future on earth . . . any failure to make this effort will make us last . . . enhance our stature . . . before this decade is out.

Kennedy continued:

Let it be clear that this is a judgment which the members of the Congress must finally make. Let it be clear that I am asking the Congress and the country to accept a firm commitment to a new course of action. . . . If we are to go only half way, or reduce our sights in the face of difficulty, in my judgment, it would be better not to go at all.

. . . It is a most important decision that we make as a nation. But all of you have lived through the last four years and have seen the significance of space. And no one can predict with certainty what the ultimate meaning will be of mastery of space.

I believe we should go to the moon, but I think

every citizen of this country, as well as the members
of the Congress, should consider the matter carefully
in making their judgment . . . as it is a heavy burden
and there is no sense in agreeing or desiring that the
United States take an affirmative position in outer space
unless we are prepared to do the work and bear the
burden to make it successful.

These last words were not in the original text but were
added by Kennedy extemporaneously, we are told by
Theodore Sorensen, the president's special assistant. "His
voice sounded urgent but a little uncertain," Sorensen recalled later, remarking that it was the only time Kennedy
had departed extensively from his prepared text in address-
ing Congress. Kennedy realized the magnitude of the task
and sensed the sustained dedication that its completion required, a dedication and national self-discipline that Americans had rarely mustered except in wartime. His concluding
remarks reflected this concern:

This decision demands a major national commitment
of scientific and technical manpower, material and
facilities, and the possibility of their diversion from
other important activities where they are already thinly
spread. It means a degree of dedication, organization
and discipline which have not always characterized our
research and development efforts. It means we cannot
afford undue work stoppages, inflated costs of materials or talent, wasteful interagency rivalries, or a high
turnover of key personnel.
New objectives and new money cannot solve these
problems. They could, in fact, aggravate them further
—unless every scientist, every engineer, every service-
man, every technician, contractor, and civil servant
gives his personal pledge that this nation will move forward, with the full speed of freedom, in the exciting
adventure of space.

The senators and representatives took notes, inspected
their fingernails, brushed their hair back and occasionally
applauded. "Where are we going to get the money?" mut-
tered one Republican as he left the crowded House floor.
Alvin Shuster of *The New York Times* reported the next
day that Congress had "embraced with some warmth" the
Kennedy objectives, but "shied at the thought of providing
all the funds to meet them."

On his drive back to the White House, Kennedy worried about the nation's response. Didn't the congressional applause seem less than enthusiastic? he asked Sorensen. Many news stories and editorials emphasized the president's foreign-aid and defense proposals. It would take a while for the people to get used to the staggering idea of men flying to the moon. In its editorial the next morning, *The New York Times* was, like many newspapers, impressed and a little overwhelmed: "The country will surely agree with the President that 'in a very real sense, it will not be one man going to the moon—it will be an entire nation, for all of us must work to put him there.'"

Congressmen remembered the national anxiety over the first countdown and the cheering crowds that had hailed Shepard's triumphant visit to Washington. They could envision the ticker-tape parades that would greet later astronauts on their return from space. More than anything, they feared the consequences of inaction. What if the Russians continued to dominate space? So, believing that the American people were willing to support an expanded long-range effort, Congress quickly and with little debate endorsed Kennedy's proposal. By August 7 it had voted $1.7 billion for NASA's fiscal 1962 budget, only $113 million less than Kennedy requested.

Thus, on the drawing boards of the space agency, in the minds of the president and his advisers and in the chambers of Congress began America's drive to send men to the moon. In 1961 no one could predict how much it would cost (estimates ran from $20 billion to $40 billion) or how long it would take (optimists talked of a 1967 landing). No one was even sure what rockets and spacecraft would be required or the flight plan they would follow. Nor was anyone sure of the ultimate value of the journey. Everyone concerned knew, however, that the task would be difficult and might be tragic—but how difficult or tragic they dared not dwell on. They were certain of one thing: it would be the greatest technological mobilization of all time.

What's in a Name?

It is interesting that in all the planning and activity hardly anyone gave a thought to naming the program. The name

Mercury for the first man-in-space project was arrived at, after high-level debate, because a Greek god had a heroic ring and Mercury was considered to be the most familiar of the Olympians to Americans—thanks more to Detroit than to the god or the planet. Later the project for two-man flights was named Gemini, naturally enough, after the constellation of the twin stars Castor and Pollux. But Apollo for the moon-landing program? When I raised the question with NASA officials a few years after, they were at a loss to explain the origin of the name. After a few telephone calls they discovered the credit should go to Abe Silverstein, who was director of flight programs in the early NASA days and went on to become director of the agency's Lewis Research Center in Cleveland. Silverstein recalled that he first used the name Apollo in January, 1959, in discussions of tentative plans for a post-Mercury program. "No specific reason for it," Silverstein said. "It was just an attractive name."

After Kennedy's announcement the name was used again, and it stuck. Apollo, the handsome son of Zeus and Leto, was the god of the sun, patron of music, maker of laws, god of healing and holiest oracle of ancient Greece. His sister Artemis (Diana), goddess of the moon, might have been a more appropriate choice. But, as an afterthought, some of the moon's program's mythologists point out that Apollo, among other things, was known as an archer who could hit targets at great distances.

Second to None

In time Project Apollo would draw its share of critics. This was inevitable. Many things can happen in a decade, and they did in the sixties. Sentiment can shift, priorities change. Kennedy's goal was seen by some as a "monumental misdecision" and the project a "moondoggle." Apollo officials would face congressional investigation of project management. There would be a steady barrage of complaints to the effect that going to the moon was an ill-advised diversion of public money from such down-to-earth endeavors as slum clearance, pollution control, education and the attack on disease and hunger.

Some of the criticism arose from a misunderstanding of

Apollo's larger purpose. True, the immediate goad was Russian competition and the specific goal was the landing of two men on the moon. But there was more to Apollo than that. Looking back on the decision, Nicholas E. Golovin, who was on the staff of the special assistant to the president for science and technology* during most of the sixties, wrote in 1967:

> . . . the underlying thread of logic leading to the Apollo decision had three strands. First, because achievement of the Apollo goal in this decade would require an unprecedented technologic and industrial effort, the main thrust of its contribution would be enhancement and broadening of U.S. space competence in general. Second, in a state of continuing competition among the great powers for world leadership, prowess in space technology had become a major qualification for such leadership . . . And, third, the manned lunar-landing goal was sufficiently difficult, challenging, and distant that its achievement by competing powers . . . could not enable any one of them to use the space program as a politically effective tool.

In short, Apollo, it was hoped, would enhance United States power and prestige, and even if the Russians beat the Americans to a manned landing they would not gain much politically so long as the Americans were not far behind.

John Kennedy, who would not live to see the goal reached, regarded Apollo "as among the most important decisions that will be made during my incumbency." According to Sorensen, Kennedy's purpose was to "provide a badly needed focus and sense of urgency for the entire space program." He said that Kennedy was "more convinced than any of his advisers that a second-rate second-place space effort was inconsistent with this country's security, with its role as a world leader and with the New Frontier spirit of discovery." Apollo offered Americans a chance to rise above complacency, to recapture their national self-confidence. This was the challenge of space in the sixties and the commitment of the Americans who set out to put mankind on the moon.

* Who was Wiesner during the Kennedy years and Donald C. Hornig during the Johnson tenure.

Chapter II

The Target

"There is no one among you, my brave colleagues, who has not seen the moon, or, at least, heard speak of it."

With those words Jules Verne's fictional Impey Barbicane, president of the Baltimore Gun Club, opened his famous address a century ago. Before that august body of restless, post-Civil War munitions men, assembled in the club's great cannon-lined hall, Barbicane proposed an expedition to seize the moon and incorporate it into the Union. This could be accomplished, he explained, by firing a manned capsule from a huge cannon implanted in the ground outside Tampa Town, in Florida.

"Three cheers for the moon!" roared the Gun Club with one voice.

All who have read Verne's *From the Earth to the Moon,* published in 1865, are struck by the French author's amazing prescience. He correctly calculated that a velocity of about 25,000 miles an hour would be required for a spaceship to escape earth's gravity in order to reach the moon. He anticipated that a test flight of animals—in Verne's book a cat and a squirrel (which the cat ate in flight)—would precede a manned launching. He even chose Florida as the launching site and the Pacific Ocean for the splashdown. He also foresaw the excitement and drama that would attend such a flight, observing that the "American public took a lively interest in the smallest details of the enterprise." In one important aspect, however, time proved Verne wrong. His American moon mission was backed by substantial Russian subsidies.

Nonetheless, the prophetic works of Verne and other science-fiction writers reflect man's long and deep fascination with the moon. The first time man saw the moon he probably reached for it. There it was night after night, sometimes full and round, sometimes a pale crescent, so mysteriously beautiful, so tantalizingly close, yet so far away that only in their fantasies did our ancestors ever fly to earth's nearest neighbor. To them the moon was, as Victor Hugo wrote, a "kingdom of dream, province of illusion."

Sphere of Fantasy

For centuries man hardly knew what to make of the moon. Was it another world of people and great civilizations? A mere reflection of earth, a benign god or goddess?

PHYSICAL FACTS ABOUT THE MOON

Diameter: *2,160 miles (about ¼ that of earth)*

Circumference: *6,790 miles (about ¼ that of earth)*

Distance from earth: *238,857 miles mean (221,463 minimum to 252,710 maximum)*

Surface temperature: *250 (sun at zenith) —280 (night)*

Surface gravity: *1/6 that of earth*

Mass: *1/100 that of earth*

Volume: *1/50 that of earth*

Lunar day and night: *14 earth days each*

Mean velocity in orbit: *2,287 miles per hour*

Escape velocity: *1.48 miles per second*

Month (period of rotation around earth): *27 days, 7 hours, 43 minutes*

The mystery inevitably bred legends and superstitions. The ancient Chinese made the moon their patron of poetry, and rare is the poet in any civilization who has not addressed the moon with great passion. Wrote John Keats: "What is there in thee, moon, that thou shouldst move my heart so potently?" To the ancient Greeks the moon was Selene, goddess of the night. (Moon geography today is called selenography.) The Romans called it Luna, whence our adjective "lunar." Hindus and most Asians worshiped a male moon deity, as did many American Indian tribes. Cherokee Indians, it is reported, behaved strangely during an eclipse. They ran wild, firing guns, beating kettles, whooping and screaming. They believed that eclipses were caused by a huge frog that gnawed at the edge of the moon and they wanted to save their revered moon by frightening the frog away.

According to various other tales:

- Ghosts walk at full moon, and ulcers as well as insanity can come from excessive exposure to moonlight.
- A red moon is an evil omen; the moon was said to be bloody at the Crucifixion.
- The dark of the moon is the time of nefarious deeds, for only then does the devil appear on earth. However, the luckiest rabbit's foot is one taken in the dark of the moon, preferably in a graveyard by a cross-eyed person.
- "The old moon in the arms of the new"—the faintly outlined disk resting on a bright new crescent—is a sign of storm. A crescent with points tilted down signifies rain, for some ancients believed the moon spilled water. A points-up crescent is a sign of a dry spell.
- A waxing moon gives energy to all living things, and so in South America to this day some mothers expose their newborn infants to direct moonlight. By contrast, the waning moon is often associated with dwindling life, or death itself.

As for the man in the moon, he was believed to have been banished from earth for stealing or cursing or breaking the Sabbath. In German legend, it was for stealing cabbages on Christmas Eve. French folklore had him Judas Iscariot, consigned to the moon as punishment for his betrayal of Jesus Christ. Dante, in his *Divine Comedy*, alludes to the tale that the man in the moon is Cain, bearing a bundle of thorns.

Until the last century some people still speculated that the moon was inhabited. About 100 A.D. Plutarch wrote *On the Face That Can Be Seen in the Orb of the Moon*, in which he described the moon as a second earth, with demons living in caves. In his *True History*, Lucian of Samosata, a second-century Greek satirist and a forerunner of modern science fiction, wrote of a voyage to the moon and the discovery of lunar inhabitants who rode about on the backs of three-headed vultures. And no less a person than Johannes Kepler, the seventeenth-century German astronomer, wrote of tough-hided lunar creatures who lived in caves to escape the sun's blistering heat. To be fair, however, Kepler's employment of exaggerated fiction in his book *Somnium* (Dream) was his indirect way of espousing the new and highly controversial Copernican science without angering church authorities, who at the time suppressed all theories that displaced earth from the center of the universe.

Sphere of Fact

By the time the Apollo astronauts began preparing for their first lunar voyage, scientists had sorted out fact from fantasy. They had gathered a great mass of data by mathematical calculation, telescope and radar.

They knew that the moon and earth make an interesting pair as they swing one around the other and the two of them, in gravitational embrace, move around the sun. The earth-moon combination moves very fast in its path around the sun—about 18.5 miles a second. The moon moves much more slowly around the earth—about 0.6 miles a second. One way to picture it is to think of the earth and the moon as a couple of dancers with arms locked, spinning wildly around a dance floor with the sun in the center. The plane in which they spin is tipped slightly, about five degrees off level. And since the earth is 81 times more massive than its partner, it is the anchorman in the dance.

Though both are globes, they are not quite as completely spherical as once supposed. Earth, a bit pear-shaped, spins, or rotates, on its axis once nearly every 24 hours—actually 23 hours 56 minutes. It makes a complete trip, or revolution, around the sun about every 365 days. This dictates the length of our days and years.

The moon is earth's captive partner. Every 27⅓ days the moon travels approximately 1,396,000 miles in an orbit around the earth. It is not a perfectly circular orbit, which accounts for the fact that during each orbit the distance between the two bodies ranges from as close as 221,463 miles to as far as 252,710 miles. Simultaneously the moon is rotating on its axis at a speed of about 10.35 miles an hour, making a complete rotation every 27⅓ days. Thus the moon rotates once on its axis in the same amount of time that it takes to make a complete trip around the earth.

This comes about because the gravitational pull of the earth controls the rotation of the moon. The result is that the moon always presents the same face to earth, while the opposite hemisphere (the so-called back side) is permanently averted. Astronomers have a down-to-earth way of illustrating this. Stand in a room, they suggest, and place a chair in front of you. You are the moon; the chair is earth. Go all the way around the chair, always facing the same wall. You revolved, but you did not rotate. Now go around the chair, facing the chair at all times. You will find that you revolved once around the chair, and you rotated once. The same half of your body was always facing the chair. So it is with the moon and earth.

Other planets in the solar system have natural satellites, too. Jupiter, the largest of the nine planets, has 12 bodies circling it; they range in size from one only 14 miles wide to the massive Ganymede, which has a diameter of 3,100 miles and is the largest known satellite. Saturn has 10 orbiting companions, including the 3,000-mile-wide Titan, as well as mysteriously beautiful rings composed of whirling particles no bigger than grains of sand. Uranus has five moons; Neptune and Mars two each, and Mercury and Venus none. If faraway Pluto has any satellites, they have gone unobserved. In fact, many astronomers speculate that Pluto may have once been a moon of Neptune until it escaped into a solar orbit of its own.

The relationship of earth to its moon, however, is in one sense unique: no other planet has a satellite that is so large in proportion to its own mass. The moon's diameter of 2,160 miles is a quarter of the earth's and almost as big as Mercury's. Consequently, astronomers often look upon the earth-moon system as a double planet.

Each of the bodies is constantly affected by the other's presence. The tidal forces between the two, for instance, are unmatched elsewhere in the solar system. The same

lunar gravitational force that causes earth's oceans to rise
and fall has also helped distort the earth into a shape
similar to that of a symmetrical egg. We do not notice it,
but lunar gravity also makes the earth's crust expand and
contract ever so slightly. And earth's constant tugging has
helped bring about a decided bulge on the side of the
moon that always faces earth.

"Sublime Desolation"

The moon's general features first came into focus when
Galileo pointed his "optick stick" at our satellite in 1609.
The surface, Galileo reported, "is uneven, rough, replete
with cavities and packed with protruding eminences."
Gradually, other astronomers filled in the details, and one
fact became overwhelmingly clear to them: men who
visited the moon would find it, in the words of Harvard
astronomer Fred L. Whipple, a "sublime desolation."

The moon has no detectable air, and, therefore, no
weather—no wind, no clouds, no rain, no snow, no water
on the surface. Men would obviously have to take along
their own air and water. There would be no life awaiting
them on the moon. Patrick Moore, a British lunar astronomer, puts it bluntly: "We have found that this 'moon-man'
of countless story-tellers does not exist; that animals and
insects are also out of the question; and that terrestrial-type
plants are certainly absent. On the whole moon there is no
living thing, apart perhaps from a few scattered patches of
lichens or moss-type vegetation on the floors of some of
the craters."

The moon's lack of atmosphere produces strange effects.
There is no sound. There is no color in the moon's sky,
only blackness, both day and night. (The blue of earth's
sky is caused by sunlight being scattered by atmospheric
gases.) By day a whitish sun lights the surface of the
moon. Day and night, stars shine in the black sky much
more brightly than when seen from earth and without the
twinkling effect produced by earth's distorting atmosphere.
One striking sight is the earth, the most prominent object
in the lunar sky. Since the earth is larger than the moon
and a better reflector of sunlight, earthshine is more than
eight times as bright as moonshine. Like the moon to us,

the earth would appear to lunar observers in phases: new earth, first-quarter earth, full earth and last-quarter earth.

Men visiting the moon would miss the protective features of an atmosphere. In the absence of insulating air, the sun beats down relentlessly during the lunar daytime. Surface temperatures where the sun is directly overhead are estimated to rise to about 250° Fahrenheit, far above the boiling point of water (212°). During the lunar night the temperatures plunge to at least 280° below zero, low enough to freeze solid the mercury in a thermometer. (On earth the atmosphere acts as a blanket, retaining much of the day's heat when the sun goes down.) Visitors must also bear in mind that, because of the moon's slower rotation, lunar days and nights each last about two weeks of earth time.

Neither is there any atmospheric protection against the occasional showers of micrometeoroids and the more dangerous bombardment of solar radiation. Scientists doubt that micrometeoroids, tiny bits of matter that fly through space, would pose a threat to properly suited astronauts. Larger meteorites hit the moon so infrequently as not to worry potential explorers. No one knows when the last meteorite struck the moon. But as early as Kepler's *Somnium,* solar radiation was discussed as a grave danger in space flight. Such radiation strikes the moon at full strength with doses that at times could prove fatal. An intense solar outburst would force astronauts to repair to a shielded spacecraft or caves or other shelters on the moon.

Whether or not the moon ever had any significant atmosphere is a question still debated by scientists. If it did, the moon's weak gravity apparently could not hold onto the gases and let them slip away into space. The moon's surface gravity is estimated at about one-sixth that of earth.

This has its advantages. For one thing, it would be easier to launch a rocket off the moon than off earth. The moon's grip on objects is not as strong as earth's. This makes it possible for astronauts on the moon to lift heavier objects and take longer, springier steps than they could on earth. A person or object weighing 180 pounds on earth would weigh only 30 pounds on the moon. Pole vaults of 60 feet would be ordinary.

Anyone who has looked at the moon can see that it has two kinds of surface—light and dark. The light shows up in telescopes as mountains and rugged highlands. The dark appears to be relatively smooth plains, which astronomers of Galileo's time mistook for seas and which on early maps were consequently labeled *maria*, the plural of *mare*, the Latin word for "sea." Some were given such names as Mare Tranquillitatis (Sea of Tranquillity), Mare Serenitatis (Sea of Serenity) and Mare Foecunditatis (Sea of Fertility). Other prominent plains include Mare Imbrium (Sea of Showers), Sinus Medii (Central Bay) and Oceanus Procellarum (Ocean of Storms), the largest of the *maria* on the side of the moon facing earth. These names have survived the knowledge that the so-called seas are actually very dry stretches of land.

Scientists have offered varied theories to explain how *maria* came to be. As recently as 1966, Dr. John J. Gilvarry of NASA's Ames Research Center argued that the *maria* are dark with the residue of life that once existed on the moon. He believed that the moon had true seas when it was young but that they evaporated and diffused into space because of the moon's weak gravity. But the seas, he contended, were there long enough for life to evolve and lay down organic sediments.

A second theory gave Apollo planners something to worry about for several years. In 1955 Thomas Gold, then an astronomer of the Royal Greenwich Observatory in England and now at Cornell University, suggested that the *maria* were covered with a layer of dust so deep that anyone who attempted to land in one might be swallowed up. This layer of dust would be the accumulation of fine particles produced by meteorite impacts and micrometeoroid showers. To Gold and his followers, one clue was the way the moon absorbed and radiated the sun's heat, suggesting some type of insulation such as dust. Astronomers, including Gold, hastened to point out that the dust might not be loose and powdery, for in a vacuum fine particles tend to stick together like clods in a freshly plowed field.

A greater number of lunar scientists subscribe to a theory that the *maria* originated as spreading pools of molten lava. According to Gerard P. Kuiper, a University of Arizona astronomer and one of the theory's leading exponents, the lava may have been produced by the enormous heat of meteorite impacts. The hypothesis is that the lava formed beneath the surface, boiled through cracks to the top and spread across the landscape, filling depressions, covering old craters and obliterating all but the loftiest mountain peaks. This would have left the fairly smooth plains that exist today. As cooling set in, a porous crust may have formed a lid on the lava. The dust partisans point out that meteorite impacts were more likely to pulverize rock than to melt it. To many astronomers the classic example of the lava process is the Sea of Showers, Mare Imbrium. They believe it was formed when an asteroid about 100 miles in diameter crashed into the moon, shaking the entire moon and melting vast amounts of subsurface material to form lava.

An even more intense debate ranges over the lunar craters, the most striking features of the moon's surface.

Thousands of craters, ranging in diameter from the 160-mile-wide Clavius down to those no bigger across than a silver dollar, give the moon its scarred appearance. A person standing in some of the broader craters would be unable to see the walls because the moon, being smaller than earth, has a surface that curves more sharply. The depth of some lunar craters is staggering. Newton* is so deep (29,000 feet) that Mount Everest would barely peep out at the top. No sunlight or earthshine ever casts a cheery ray on Newton's eternally frozen floor. A number of craters, such as the 75-mile-wide Alphonsus, are a puzzle to astronomers because of the mountain peaks rising from their floors. The peaks are generally near the center and apparently never rise as high as the surrounding walls. There are even craters within craters, as in the case of the famous Copernicus, the so-called "monarch of the moon." Typical of many "young" craters is Tycho Brahe, a 56-mile-wide depression in the south of the moon. Out from its rim extend radiating streaks of very light materials, some for 1,000 miles. The sunburst effect makes Tycho one of the most brilliant sights on the moon.

Prevailing scientific opinion has it that most lunar craters were formed by the impact of meteorites and comets. However, the case is not closed, and a minority insists that the craters resulted from volcanic eruptions.

Advocates of the impact theory believe that when the

---

* In 1651 the Italian astronomer Giovanni Riccioli adopted the system of naming the craters after notable personalities, usually scientists. The practice is still followed.

moon was young it was a time of violent disturbances in space. The interplanetary regions were littered with rocky debris moving at high speeds. The exploding impacts of such objects would have scattered lunar soil for great distances, possibly explaining the rays, and piled mounds of materials on the craters' edges, explaining their high rims. Similar fragments no doubt hit earth at the same time, but the erosion of water and wind smoothed out the scars.

This, however, does not rule out some volcanic activity on the moon. "There is growing evidence," says John W. Salisbury, chief of the lunar-planetary branch of the air force's Cambridge Research Laboratories, "that both processes have produced craters on the lunar surface."

Other features of the landscape include the highlands and mountains, fills and ridges, deep faults and long valleys.

In the highlands, believed to be the oldest parts of the moon's surface, the craters are so numerous that they overlap. Their debris seems to have piled up to form the surrounding hills, and some of the ridges are thought to be remnant walls of old craters. From the edges of the *maria* rise chains of mountains. The largest, the Leibniz range, which lies near the south pole, has peaks up to 30,000 feet.

The sinuous rills, another common feature, are streaks or cracks that look like meandering streams and were once viewed as evidence of water existing on the moon in past times. Actually their origin is not clear. Some astronomers speculate that they were "riverbeds" for hot volatile gases and pulverized lava escaping from fissures. Deeper breaks, or faults, in the lunar surface have also been observed, indicating the occurrence of moonquakes from time to time.

Though it is popular to speak of the moon as a dead world, an orbiting cinder, scientists are not so sure. The craters Aristarchus and Alphonsus cause them to wonder. In 1956, a Russian astronomer, Nikolai A. Kozyrev, trained the Crimean Astrophysical Observatory's 50-inch telescope on Aristarchus and the spectrograms revealed what appeared to be gases emanating from the lunar crust. It was not conclusive. But in 1958, while investigating the peak in the center of Alphonsus, Kozyrev registered a half-hour release of gas, suggesting some stirrings of molecules in the otherwise desolate moon.

Origin of the Moon

Until this century few scientists bothered to seek an answer to a fundamental question about the moon: Where did it come from? They may have been constrained by biblical teachings as to the creation of the earth and merely assumed that the moon also was a product of divine intervention. In any case, there are now three scientific hypotheses in search of confirmation by lunar explorers.

In the first, the moon is thought to have formed somewhere in space in which it wandered until captured by the earth's gravitational field. The second holds that the moon was once part of earth, then split off but could not entirely escape. In the third, the moon is believed to have evolved as a separate body at the same time as earth, perhaps from materials left over from the formation of earth. There is general agreement, however, that the moon and the earth were formed at approximately the same time. That would make the moon about 5 billion years old.

It was Sir George Darwin, the astronomer son of the famous naturalist, who in 1898 proposed one of the first theories. On the basis of a study of gravitational forces, he suggested that the Pacific Ocean could be a great hole torn in the earth's relatively thin crust by a cataclysm that ejected the moon. According to Darwin, the young earth was a molten sphere spinning very rapidly so that the flattening effect of this faster and faster rotation, together with the strong gravitational pull of the sun, could have flung a huge blob of earth off to form the moon. Interesting as the theory is, most scientists came to dismiss it as impossible—until, that is, the Space Age revived the speculation.

Another conjecture is sometimes called the "hot" theory, which is related to one of the older ideas of planetary creation. It is based on the assumption that a passing star tore great masses of material from the sun and as the material condensed and cooled it formed the planets and their satellites. Thus the moon would have been created the same way as earth and at roughly the same time. In this view, as in Darwin's, the moon's composition would be much the same as earth's.

A more recent theory was offered in 1943 by C. F. von Weizsäcker, the German astrophysicist. He suggested that the planets and their satellites condensed from whirlpools of interstellar dust and gas left over after the sun was formed. A mechanism of this kind—and some form of it has come to be the most generally accepted theory—can explain how most of the spin goes into the planets rather than the sun. In 1949 Kuiper expanded on the theory to describe how the earth-moon combination may have come about. Out of these clouds of dust and gas, Kuiper theorized, came large separate revolving clouds called "protoplanets." At this stage the protoplanet representing the earth split into two parts: a larger cloud, which became earth, and a smaller protomoon coupled to it. They eventually condensed into their present forms.

Whatever the origin of the moon the prospect of men traveling to it generated during the 1960s a revival of scientific interest in earth's nearest neighbor. To many astronomers the moon had seemed a dull, changeless and uninteresting place and fairly unrewarding of study compared with Mars or Jupiter or the distant stars. The Apollo project changed this attitude. Scientists—astronomers, geologists and physicists—began to respond to the enthusiasm of Nobel laureate Harold C. Urey of the University of California at La Jolla. Urey saw in the moon's very changelessness an opportunity to answer some of the fundamental questions about the solar system. In short, the moon could yield the answer to how the earth was created. "Because its surface has preserved the record of ancient events," says Robert Jastrow of NASA's Goddard Space Institute, "it holds a key to the history of the solar system."

In the summer of 1965, a group of distinguished scientists from the National Academy of Sciences met at Woods Hole on Cape Cod, a place where men usually spend their time exploring the inner space of the oceans, and considered what the first men on the moon should do to gather as much scientific information as possible in a short time. The scientists proposed 15 major questions as guidelines for lunar exploration. They wanted to know the moon's internal structure, the surface composition, the principal processes responsible for the structure of the lunar surface, the age of the moon, any traces of volcanic activity, the rate of solid objects striking the moon, the history of cosmic and solar radiation acting on the moon, and the presence of any magnetic fields retained in lunar rocks.

Not all the questions could be answered on the first landing, to be sure, but the instruments the men carried to the moon should be designed to probe for such answers. The suggestions included a seismometer to detect any tremors beneath the moon's surface, either from a moonquake, volcanic activity or the reverberations from a meteorite impact. The scientists also wanted the astronauts to return as many samples of lunar soil as possible. They suggested an experiment to entrap any gases spewed out by the sun to bombard the moon. They had other, more ambitious suggestions, but NASA finally decided that setting up a seismometer, collecting rocks and laying out a gas detector and other instruments would occupy enough of the time and energy of the first men on the moon.

With these scientific goals in mind, in addition to patriotic pride and engineering challenge, the United States began preparing a fleet of unmanned picture-taking spacecraft to scout out the moon in advance of the men of Apollo. These were the Ranger crash-landers, Surveyor soft-landers and Lunar Orbiters (see Chapter V). Besides transmitting a wealth of intriguing scientific data, these robots would reassure Apollo planners that they knew how to get to the moon.

Chapter III

How to Get There

For more than a year after President Kennedy's decision in 1961 to try for the moon, a fierce debate raged among top advisers of the Apollo project over what would be the safest, simplest and most practical way to fly men to the moon, and achieve it in the decade. Once, the debate even exploded into a shouting match in front of the president. About the only points some of the Apollo planners could agree on for months were the fundamental laws of nature that govern man's flights into space, just as they govern the earth's flight around the sun and the moon's flight around earth. No one could dispute Kepler or Newton and hope to reach the moon.

Man's intellectual preparations for a voyage to the moon began back in the seventeenth century when the great German astronomer Johannes Kepler pondered the movement of planets, and when the English genius Isaac Newton supposedly contemplated the falling of an apple. They, in turn, owed a debt to the Polish monk Nicolaus Copernicus who, in the previous century, had broken with a thousand years of traditional thought to propound the theory that the earth was not a fixed body at the center of the universe, but one of many planets moving around the sun. This theory led to modern man's conception of a solar system in which all the planets, the earth included, are in continuous motion.

But it was the brilliant insights of Kepler and Newton into the workings of the universe that explained these planetary motions and provided the theoretical underpinning for the Space Age. Thus, to understand the principles of rocketry, the orbits spacecraft follow, or the forces that keep a satellite up, one must first understand the laws of Kepler and Newton.

Bodies in Orbit

Astronomers before Kepler—and even Kepler at first—had attempted to explain planetary motions in terms of circles. The circle was considered to be noble in its perfection and thus appropriate to an orderly universe. Using detailed observations made by the Danish astronomer Tycho Brahe, Kepler calculated that a planet does not travel in an exact circular orbit around the sun, but in a slightly elliptical one, with the sun at one focus of the ellipse. This became known as Kepler's first law, which he enunciated in 1609. The law applies equally to a lesser body than a planet—to the moon or a man-made satellite. For a satellite of the earth, the center of mass of the earth, not the sun, acts as one focus of the ellipse. True to the law of ellipticity, the moon's orbit varies between 221,463 miles to 252,710 miles out from earth. Knowing this condition helped make it possible for men to determine long in advance where the moon would be at any given time astronauts wanted to embark for a landing.

Kepler also recognized, in his second law, that as a satellite moves in its orbit, the line joining the satellite's center to the focal point (the body around which it is orbiting) will always sweep out equal areas in equal periods of time. To do so, an object in elliptical orbit must travel faster when it is closer to the body it is orbiting than it does when it is farther away. According to this principle, a planet or spacecraft loses speed as it rises in orbit. As an example, take Pluto, whose orbit is the most eccentric of the sun's planets: Pluto's speed is 3.8 miles a second at its closest approach to the sun and 2.3 miles a second at its farthest point.

Kepler's third law, called the harmonic law, deals with the relationships between the distances of the planets from the sun and the time required for one full revolution around the sun. When a satellite, natural or man-made, makes a full trip in orbit around its primary body, it is said to complete a revolution, and the time required is termed its period, or period of revolution. Kepler was the first to state that an orbiting body's period is proportional to its mean distance from the primary focus of its orbit. Thus, the higher the orbital path is from the center of the body it is orbiting, the longer the orbital period.

For instance, the early man-made satellites usually

traveled little higher than 100 miles above earth (about 4,000 miles from its center). At that altitude, they traveled about 17,400 miles an hour and took no more than 90 minutes to circle earth. By comparison, the moon, at a mean altitude of 239,000 miles, is traveling only about 2,287 miles an hour and thus takes 27⅓ days to complete a full revolution of earth.

Inescapable Gravity

Although Kepler was able to describe accurately the motions of the planets, he did not arrive at the answer as to what force keeps the planets (or any satellite) moving as they do in their orbits. It was Isaac Newton who did this in formulating the law of universal gravitation in 1666, and the three laws of motion in 1687. These laws explain why Kepler's observations were correct.

The law of universal gravitation states that every body in the universe attracts every other body with a force that is directly proportional to the product of the masses of the two bodies, and is inversely proportional to the square of the distance between the two bodies. This force is what we call gravity. Like magnetism, it is invisible. Its essential nature and source are unknown. But its effects are well known. Gravity is what causes the apple to fall to the ground. It keeps man, under ordinary circumstances, rooted to the earth's surface. It causes bodies in the solar system—the sun and the planets, the planets and their satellites—to exert a mutual pull on each other.

For the purposes of spaceflight, the important point made by Newton is that the pull of earth's gravity falls off steadily as distance from the earth increases. Therefore, the greatest burst of power is required to make the initial escape from the earth. After that, a little rocket power will take a spacecraft a longer way.

This diminishing pull of gravity must be carefully considered in a flight to the moon. To leave the vicinity of earth a spacecraft must rocket hard enough to get to the moon, but not hard enough to break the bonds of earth's gravity altogether. A spacecraft, it has been learned, escapes from the earth's gravitational field when it reaches a speed of about 26,000 miles an hour. If a spacecraft reached that speed, and then missed the gravitational field of the moon, it would keep on going into deep space and swing into orbit around the sun.

For a spacecraft to leave the vicinity of the earth and head toward the moon, in what is really a highly elliptical earth orbit, requires a speed of about 24,200 miles an hour. This is fast enough to pull away from earth's strongest gravitational grip but slow enough to remain under the earth's influence. Then at a point about 38,900 miles from the moon, the spacecraft enters the moon's sphere of gravitational influence. Gravity is still at work, but it is the gravity from a different body that is strongest. Because the moon's gravity is weaker than that of earth, a spacecraft's speed in lunar orbit is slower than it would be in earth orbit. For the same reason, it takes less power to blast off the moon's surface and escape lunar gravity than to push off from the earth's surface.

What Keeps a Satellite Up?

Newton's three laws of motion went further in explaining planetary and satellite behavior. The first law states that a body that is at rest (or in motion) will remain at rest (or will move in a straight line) unless acted on by some external force. This property is called inertia.

When a spacecraft is launched, it wants to travel in a straight line. This would keep taking the spacecraft away from the surface of the earth. A rifle bullet or cannon shot also starts out in a straight line, but soon gravity and atmospheric friction pull the shot, arclike, back to the ground. However, if a spacecraft is launched with sufficient velocity (some 17,400 miles an hour), it will not plunge back to earth but will go into orbit, where the pull of gravity balances the tendency to fly off straight into space. The result is a compromise: a state of equilibrium between the forces of inertia, derived from the rocket that pushed the spacecraft into space, and of earth's gravity. Inertia keeps gravity from winning, and gravity keeps inertia from winning. The balance keeps the spacecraft in orbit. Its path curves as though it would fall back to earth. But at the same time, the surface of the earth itself, being round, falls away from the satellite just as fast as the satellite falls. Consequently, the satellite "falls" around the earth, never reaching the surface.

What is called zero gravity, or weightlessness, the familiar effect that causes any object loose in an orbiting spacecraft to float about, has nothing to do with being beyond the influence of gravity. It is the result of the forces of inertia and gravity canceling each other out.

As long as the satellite maintains the same velocity, it will continue to orbit earth at the same altitude. On earth a body in motion eventually comes to rest because of friction with the air. In space, well above the atmosphere, there is no appreciable friction. To return to earth the spacecraft must slow down through the braking action of a rocket.

But there is something called orbital decay. At the relatively low orbital altitudes of about 100 miles, there are still a few air molecules to bump into and a fairly strong tug of gravity. After a spacecraft bumps into enough of these molecules, it slows down enough so that gravity will win out and pull the spacecraft back toward earth. The result is that satellites at such low altitudes stay up only a few weeks. Those as high as 1,000 miles can stay up for many years, and, of course, the moon has been up there in orbit for millions of years.

One of the facts that becomes evident from the laws of Kepler and Newton is that for every altitude above the earth, there is a precise orbital velocity. As Kepler discovered, the higher the path, the slower the satellite moves and the longer it takes to complete one revolution. If a satellite is injected into orbit with a greater velocity than is required for a perfectly circular orbit at that altitude, the orbit of the satellite will be one of Kepler's ellipses. The precision of velocity necessary for a circular orbit is all but unattainable; this explains why the orbits of the planets and the earth satellites are all ellipses.

These laws also explain the seeming paradoxes of orbital flight. To attain a faster orbiting speed, you slow down, so to speak. To lose speed, you accelerate. In other words, if a spacecraft wants to go faster it must fire its braking rocket to change the equilibrium between gravity and inertia. Gravity thus pulls the spacecraft into a lower orbit where, being closer to earth, the ship (according to Kepler's laws) goes faster than in the previous higher orbit. But if a spacecraft fires its rockets to speed up, it attains a velocity sufficient to get into a higher orbit and consequently (again according to Kepler) is traveling at a slower speed.

Newton's second law of motion states that a force acting on a body causes it to accelerate in the direction of the force, the acceleration being directly proportional to the force and inversely proportional to the mass of the body. This means that the greater the mass of the body you want to move, the greater the force required to move it—and once it is in motion, the greater the force that is required to slow or stop it. In modern space terms, it takes a more powerful rocket to boost 100,000 pounds toward the moon (as in the case of Apollo) than a few hundred pounds into a low earth orbit. The same principle applies to designing braking rockets to slow the spacecraft so that it can land on the moon.

What Gives a Rocket Its Push?

The third and best known of the Newtonian laws of motion states, simply, that for every action there is an equal and opposite reaction. If this were not true, there would be no recoil from a rifle shot and rocket engines would not work at all.

The force that propels rockets, called thrust, has often been demonstrated with an ordinary toy balloon. If you suddenly release a blown-up balloon, the air inside will rush out the balloon's open neck. Obeying Newton's action-reaction law, the rushing air creates a reaction force that drives the balloon in the opposite direction. Just as a balloon is thrust forward by expelling air backward, a rocket is thrust forward by expelling particles, usually in the form of a gas, from its rear nozzle. The greater the flow rate through the nozzle, the greater the forward thrust.

An important fact is that the forward motion of the balloon is not caused by the air expelled pushing against the atmosphere. Even if there were no atmosphere to push against, the balloon would still zip around the room. In fact, it would go farther and faster. A rocket engine works more efficiently at higher altitudes and in space than it does near the surface of the earth. In those higher regions, the atmosphere is either very thin or the vehicle is operating in what is essentially a vacuum. The rocket must, therefore, take along its own oxygen to support combustion.

But it is worth the trouble. In space, the gases of the rocket exhaust do not have to push against the pressure exerted by the molecules of air. They can escape from the engine much faster than they can close to earth and, consequently, can produce a greater action and a greater forward reaction, or thrust.

A rocket's thrust is measured in pounds. To launch a rocket from the ground, the number of pounds of its thrust must be greater than the number of pounds it weighs; it must have more push than it has weight. This involves another tug-of-war, between earth's gravity and the rocket engine's power. If the engine can push up (thrust) more than gravity is pulling down (weight), the rocket will move up. How rapidly the rocket moves up depends on how much greater its thrust is compared with its weight. Rocket engineers call this the thrust-to-weight ratio.*

In a broader sense, however, Newton's contribution to the Space Age was his mathematical demonstration that planets, or any orbiting bodies, obey the same physical laws of motion and gravity as bodies on the earth. It made the vastness of space somewhat less strange and, by analogy with the behavior of matter on earth, more understandable.

From Principle to Practice

The world of Kepler and Newton was, of course, a long way from spaceflight. Man got about by horse and by sail, and only in his dreams left the earth for the orbiting worlds beyond. But the new discoveries of astronomy and the emerging spirit of invention that was to lead to the Industrial Age prompted the more imaginative minds to devise fictional ways for man to move out into space, usually to the moon.

Science-fiction writers suggested such means of getting there as chariots drawn by swans, ladders, balloons, cannon and capsules coated with some sort of gravity-defying sub-

---

* Apollo's Saturn 5 rocket, together with its Apollo spacecraft, weighed slightly more than 6 million pounds. The thrust of the first-stage engines, which at lift-off had to labor against the strongest pull of gravity, was more than 7.5 million pounds, enabling the rocket to win the tug-of-war.

stances. Cyrano de Bergerac deserves credit for one of the more original, if equally impractical, ideas. Observing that the dew was drawn into the sky by the morning sun's rays, he hit upon his idea. "I planted my selfe," Cyrano wrote, "in the middle of a great many Glasses full of Dew, tied fast about me; upon which the Sun so violently darted his rays, that the Heat, which attracted them, as it doese the thickest Clouds, carried me up so high, that at length I found my selfe about the middle region of the Air."

The method left much to be desired, even in fiction. Cyrano landed in Canada, not on the moon. But he then had a more practical idea. He attached a box containing a number of large firecrackers and exploded them to head for the moon. Without realizing it, Cyrano was apparently the first to guess the proper principle for moonflight, the principle of reaction. This took place half a century before Newton arrived at his laws.

The history of rocketry goes back long before Newton or Cyrano to ancient China—how far back, no one seems to know—and the invention of the ordinary skyrocket. It was more the ancestor of the Fourth of July fireworks than of modern rockets, but the same principle was being put into practice. "Arrows of flying fire," as the military versions were called, went into use as early as the thirteenth century by the Chinese against the invading Mongols. And who can ignore the intrepid Wan-Hoo? As the story goes, and it is probably apocryphal, a Chinese official named Wan-Hoo assembled two large kites with a saddle and 47 powder rockets. Seating himself in the saddle, he signaled 47 waiting coolies to ignite the rockets with flaming torches. Wan-Hoo, alas, succeeded only in killing himself in an explosive cloud of black smoke.

Europeans came up with larger rockets in the early nineteenth century. More than 25,000 rockets developed by William Congreve were launched by the British against Copenhagen in 1807. Similar rockets, weighing about 30 pounds each, were also used in the War of 1812, inspiring Francis Scott Key's words "the rockets' red glare," in "The Star-spangled Banner."

Not until the turn of this century, however, did anyone recognize fully the value of the rocket for space travel. That it took so long is understandable. Who in an earlier time took space travel seriously? (How many, for that matter, took it seriously before Sputnik?) And who could picture anything related to those crude and erratic Congreve rockets being aimed at the moon? No one holding professorial rank at a famous seat of learning could be bothered with such visionary schemes. The idea that the rocket was key to space travel grew instead in the mind of a shy, deaf and largely self-taught schoolteacher in Russia, a lonely man named Konstantin Eduardovich Tsiolkovski.

Tsiolkovski the Prophet

This man who has come to be known as the "father of astronautics"—Tsiolkovski—was born in September, 1857, almost exactly 100 years before the launching of the first spacecraft. Working with meager resources in Kaluga province far from the mainstream of scientific thought, he became the first man to work out the mathematical calculations of spaceflight. His pioneering papers were written in the 1880s and were first published in 1903, the year of the Wright brothers' airplane flight.

"For a long time," Tsiolkovski wrote, "I thought of the rocket as everybody else did—just as a means of diversion and of petty everyday uses. I do not remember exactly what prompted me to make calculations of its motions. Probably the first seeds of the idea were sown by that great fantastic author Jules Verne—he directed my thought along certain channels, then came a desire, and after that, the work of the mind."

Tsiolkovski anticipated and solved theoretically almost all the important engineering difficulties of spaceflight. Not only did he suggest the rocket as the means of escaping earth's atmosphere, but he observed that liquid fuels would provide a higher exhaust speed (hence greater reaction) than the powders and solids being used in rockets of his day. He proposed kerosene (which was used in the first stage of the Saturn 5) and later noted the possible use of liquid oxygen and liquid hydrogen, which would become the most sophisticated of the propellants in the 1960s. In addition, Tsiolkovski made detailed drawings of rocket combustion chambers and nozzles and analyzed the value of the multistage rocket, which he called the "rocket train." As each stage consumed its propellants, the Russian noted, it would be discarded to keep the weight of the vehicle as low as possible. Nearly all modern rockets have two or three stages for just that reason.

In time Tsiolkovski turned from lonely scholar to romantic prophet. Besides his dozens of scientific papers, he wrote a number of science-fiction novels. Before he died in 1935, he told of a dream of his in which new airlines were being inaugurated from Moscow to the moon and from his home town to the planet Mars. The obelisk over his grave at Kaluga bears his vision of man's destiny in the Space Age:

> Man will not stay on earth forever, but in the pursuit of light and space will first emerge timidly from the bounds of the atmosphere and then advance until he has conquered the whole of circumsolar space.

Goddard the Experimenter

Tsiolkovski never got around to performing any practical experiments based on his calculations and drawings. He was the theoretician, and something of the poet of early astronautics. The man who first combined theory and practice was an American, Robert Hutchings Goddard, who is called the "father of modern rocketry."

He, too, was a lonely pursuer of a dream who went largely unnoticed by contemporaries. Goddard was by nature a cautious man, sensitive to ridicule, but quietly determined. Born in 1882 in Worcester, Massachusetts, he grew up on the science fiction of H. G. Wells and Jules Verne and wrote of navigation in space while in high school. By the time he received his doctorate from Clark University in his home town and began teaching physics there, Goddard had discovered, independently of Tsiolkovski, that liquid oxygen and liquid hydrogen would make highly efficient rocket propellants. In 1919, he wrote a learned exposition on rocketry as a method of "reaching extreme altitudes"—space.

Shortly thereafter, Goddard came to the attention of newspapers when he suggested sending an unmanned rocket to the moon. They began calling him the "moon man." *The New York Times* printed an editorial-page article in January, 1920, suggesting that Goddard was either playing a joke or was ignorant of elementary physics if he thought that a rocket could work in a vacuum. The public's scoffing turned Goddard into a loner for life.

Goddard kept busy, though. In March, 1926, on the farm of his Aunt Effie near Auburn, Massachusetts, Goddard launched the world's first liquid-propellant rocket. A modest 12-foot projectile, it reached a height of only 41 feet, a distance of 184 feet and a speed of 60 miles an hour.

As his rockets grew bigger and the launchings more frequent, Goddard was forced to abandon the Massachusetts farm for a ranch in the wide-open spaces of New Mexico. With a $50,000 grant from the philanthropist Daniel Guggenheim, arranged by the great aviator Charles Lindbergh, Goddard, his wife and four assistants set up their workshop and launching stands near Roswell in 1930. For more than a decade they conducted a series of increasingly more sophisticated rocket experiments, culminating in flights of up to 9,000 feet and of speeds approaching the supersonic.

Goddard was awarded more than 200 patents based on his experimental work. He was the first to develop lightweight pumps suitable for rocket fuels and to blueprint self-cooling rocket motors and variable-thrust engines. He was the first to prove experimentally that (no matter what the *Times* editorial writer might think) a rocket will provide thrust in a vacuum.

A few months before his death in 1945 Goddard had a chance to inspect a captured German V-2 rocket engine. He viewed its pumps and valves and intricate tubing. "Everything," Goddard said to an assistant, "is taken from our work in Massachusetts and New Mexico." And Wernher von Braun, after getting a look at Goddard's drawings and patents that predated the V-2 development, is reported to have said, "Goddard had everything."

Yet, during World War II, Goddard's offers to pursue rocket research for the War Department were ignored. He eventually was given a contract but in an area far removed from space and missiles. Goddard's contribution was not recognized until long after his death and after the Space Age burst to life.*

---

* One of the NASA centers, medals, awards, symposia and a library at Clark University are named in his honor.

Oberth the Promoter

The third rocket pioneer, and the only one to live to see men try to fly to the moon, was Hermann Oberth, a Hungarian-born (Transylvanian) German. Born in 1894, Oberth reached many of the same conclusions about rocketry as did Tsiolkovski and Goddard. He did it independently, for in those days there was no exchange of ideas on rocketry from country to country.

In 1923, Oberth wrote a short book in which he demonstrated that a rocket could operate in a vacuum, and he outlined the speed and fuel requirements of a spacecraft capable of traveling beyond the atmosphere. An expanded version of the book, *The Way to Space Travel,* is considered the classic of early space science writing.

Unlike Goddard, Oberth did everything he could to publicize his work and to inspire others to pursue research in rocketry. His work and writings were instrumental in the formation of the German Society for Space Travel (Verein für Raumschiffahrt). Oberth and the VfR played no small part in Germany's becoming the first nation in which rocketry attracted widespread attention and the enthusiasm of a hard core of young engineers, including a young Berlin student named Wernher von Braun.

Out of Oberth's writings and the rocket enthusiasts who gathered around the "proving ground" in Berlin grew the foundations for the well-known German rocket accomplishments of World War II. Under the direction of von Braun and German army general Walter Dornberger, the German rocket engineers got the first chance to put their experiments to work on a grand scale. The German army is said to have pumped some $100 million into the building of the rocket development and launching complex at Peenemünde on the Baltic Sea.

The rockets that came out of this effort, especially the famous V-2, did not have a profound effect on the course of the war. But they did on the future of man's thrust into space. Many of the rockets were captured at the end of the war by the Russians, who took them back to their homeland to serve as the nucleus of their postwar experiments. Von Braun and his closest associates, when they saw the end near, chose to be captured by the Americans, and were brought to the United States to continue their pioneering work.

Five Ways to the Moon

For all the theoretical and experimental foundations that had been laid, no clear consensus existed in 1961 among NASA officials and presidential science advisers on exact means of landing men on the moon. The problem was not only one of getting the explorers to the moon, but of returning them safely to earth. It was a matter of deciding how much rocket power could be harnessed, how big a payload could be landed gently on the moon and how to boost the men off the moon and back to earth. Some of the many proposals, when viewed with 20-20 hindsight, seem absurd. Air-force scientists who first seriously considered the problems of lunar landing in 1958 even examined for a time the possibility of setting men on the moon from lunar orbit using little more than a type of Buck Rogers rocket belt.

In the months following the Kennedy decision, five basic schemes for a manned round trip to the moon were evaluated. These were direct ascent, earth-orbit rendezvous, the tanker concept, lunar-surface rendezvous and lunar-orbit rendezvous.

At first the most popular method was direct ascent. It seemed to be the simplest. A monster, three-stage rocket would place a spaceship of some 150,000 pounds into a lunar trajectory. The complete spaceship would go into lunar orbit and then, with the braking rocket firing, back downward to the moon's surface. The same ship would rocket off the moon and head directly back to earth. Although many top NASA engineers liked the concept, they soon realized that the development problems were insurmountable in time for a landing in the decade. Direct ascent would involve building a rocket nearly twice as powerful as anything within sight—the so-called Nova rocket, which would have had an initial thrust of some 13 million pounds. "If we had to depend on the Nova," an aerospace corporation official said later, "we'd be waiting another ten years."

Nor was there any assurance at the time that a vehicle as large as the 150,000-pound spacecraft envisioned might not break through the lunar crust or sink in the thick dust which many scientists assumed blanketed the moon. There was also the possibility that such a tall vehicle (80 to 100 feet) would topple over upon landing. "It would have been almost like trying to land the Washington Monument on its base," a NASA engineer commented.

The second alternative was to assemble separate parts of Apollo and its rockets after they had rendezvoused in earth orbit. Von Braun originally favored this technique, and for months it appeared to be the most likely choice. This would have involved smaller rockets launching a spacecraft in as many as five pieces and assembling them while they were circling the earth; if the "Nova Junior," or Saturn 5, were made ready in time, it might have taken only two separate launchings. In either case, the weight of the assembled spacecraft would be about the same as the one for the direct-flight mode, and consequently the landing would be just as questionable. Another drawback was the split-second timing necessary for the multiple launchings.

The tanker concept was a variation of the earth-orbit rendezvous method. The idea called for sending an unmanned tanker rocket into earth orbit. Once the orbit of the tanker had been carefully determined, the large manned Apollo vehicle would be sent into orbit by a Saturn 5. The Apollo would have to rendezvous with the tanker, fuel up, then separate and, using the added fuel, rocket toward the moon. The advantage of this method would have been a considerable reduction in the lift-off weight of the manned vehicle. But transferring fuels in orbit, especially the supercold liquid oxygen, presented a problem of undetermined complexity.

Brief consideration was given to a proposal to rocket extra fuel and supplies to the moon's surface aboard unmanned vehicles—the lunar-surface rendezvous idea. After the astronauts landed, following a direct flight, they could refuel from the cached supplies before returning to earth. Proponents of this lunar-surface rendezvous method admitted that there might be no way of knowing if the supplies had landed intact. There was the added risk that the astronauts might land too far from their previously launched supplies, and would be unable to refuel. This would leave them stranded on the moon.

Through the summer and fall of 1961, as the various

factions wrangled, almost no thought was given officially to a fifth possible way to send men to the moon—lunar-orbit rendezvous, or LOR. The prime contenders were direct ascent and earth-orbit rendezvous, or EOR. NASA study committees either ignored LOR or placed it toward the bottom of their lists of choices. In the fall, James Webb, the NASA administrator, and Brainerd Holmes, NASA's newly appointed administrator for manned spaceflight, announced that EOR looked like the best method.

Voice in the Wilderness

Finally, out of angry frustration, an obscure aeronautical engineer at NASA's Langley Research Center went over the heads of his immediate bosses and in November, 1961, fired off a letter to Robert C. Seamans, Jr., NASA's associate administrator. "Somewhat as a voice in the wilderness . . ." John C. Houbolt, the 41-year-old engineer, began his letter. "I have been appalled at the thinking of individuals and committees."

Houbolt went on to outline the LOR concept that he and some Langley associates* had decided over a year earlier would most easily meet the constrictions of time, money, safety and existing technology. In Houbolt's scheme, a single Saturn 5 rocket would boost toward the moon an Apollo vehicle carrying three astronauts and a detachable lunar landing craft. The Apollo vehicles, still attached, would go into an orbit of the moon. Then two of the three astronauts would leave the Apollo command ship (with one man left aboard), crawl into the landing craft, detach it and descend to the moon. After a brief stay on the moon, the men would launch the craft to return to a rendezvous with the command ship, which would still be in lunar orbit. They would transfer back into the command ship and, leaving the landing vehicle in lunar orbit, would fire a rocket to head back to earth.

Houbolt, a Midwesterner with prematurely graying hair, was not even supposed to be working on lunar-landing

---

* C. E. Brown, A. W. Vogeley, M. C. Kurbjun, J. D. Bird, R. W. Stone, Jr., J. M. Eggleston, R. F. Brissenden, W. H. Michael, Jr., M. J. Queijo, J. A. Dodgen and W. D. Mace.

plans. As the associate chief of the flight dynamic loads division at Langley, he was responsible for the investigation that determined the structural weakness that was causing the many Electra turboprop jet crashes in the late 1950s. But in his spare time he had done a lot of thinking about moonflight and had met regularly with an informal group to discuss spacecraft designs. This grew into a full-fledged committee to study space rendezvous problems beyond Apollo. When the subject of the lunar mission came up at a committee meeting in August, 1960, Houbolt went to the blackboard (engineers seem to think better with a black-board or scratch pad handy) and listed all the conceivable ways of getting to the moon, among them the lunar-orbit rendezvous.*

It was not an original concept. The first person known to have proposed LOR was a self-educated Russian me-chanic named Yuri Vasilievich Kondratyuk, who calculated in 1916–17 that LOR was the best means of achieving a lunar landing with the maximum conservation of fuel.† When a spacecraft reaches the moon or a planet, Kon-dratyuk wrote, "the entire vehicle need not land, its velocity need only be reduced so that it moves uniformly in a circle as near as possible to the body on which the landing is to be made. The inactive part (the landing craft) separates from it, carrying the amount of the active agent (pro-pellant) necessary for landing the inactive part and for subsequently rejoining the remainder of the vehicle."

Little attention was paid to Kondratyuk's dreams, espe-cially by the Soviet government. To amplify his ideas, Kondratyuk published a book in 1929, *The Conquest of Interplanetary Space,* with his own money after Moscow denied funds for the printing. He died in obscurity in 1942. Even with the advent of the Space Age, Kondratyuk con-tinued to be ignored in his native land, for the Soviet Union seemed to concentrate on earth-orbit rendezvous in its preparations for lunar and planetary flight.

Yet, the more Houbolt and his group studied LOR, the more they were convinced it was quite feasible. In 1963,

---

* In 1963, Houbolt left NASA to become a private aerospace consultant in Princeton, N.J. He became a senior vice president of the Aeronautical Research Associates of Princeton, Inc.

† Oberth also discussed LOR possibilities in 1929. Harry E. Ross, head of the British Interplanetary Society, updated Kondratyuk's ideas in 1948.

Houbolt wrote: "I can still remember the 'back of the envelope' type of calculations I made to check that the scheme resulted in a very substantial savings in earth-boost requirements. Almost spontaneously, it became clear that lunar-orbit rendezvous offered a chain reaction simplification on all back efforts: development, testing, manufacturing, erection, countdown, flight operations, etc. All would be simplified. The thought struck my mind, 'This is fantastic. If there is any idea we have to push, it is this one!' I vowed to dedicate myself to the task."

The initial reception to Houbolt's LOR proposal was disheartening. He argued that the concept would save a great amount of fuel—enough for the entire mission to be launched by one Saturn 5. Because the lunar lander would not need a heavy heat shield for a return through earth's atmosphere and would not have to carry supplies for the full mission, it could be tens of thousands of pounds lighter than the landing vehicles conceived for EOR or direct ascent. "We proved to ourselves that LOR could be done, but nobody in the committees would listen to us," Houbolt recalled in an interview in 1969 with Richard D. Lyons of *The New York Times*. Houbolt added that "they would only consider direct ascent and earth-orbit rendezvous. It was a classic example of closed-mind thinking."

Apollo planners thought the idea too risky. Their main objection seemed to center on the task of having astronauts perform intricate rendezvous maneuvers 230,000 miles from earth. Would it not be less risky, they asked, to attempt the rendezvous in earth orbit where, if anything went wrong, it would be easier to bring the men back to earth? Some scientists were worried that the small lander for LOR would be insufficient to carry instruments for any significant study of the moon.

At a meeting of Apollo planners in Washington in early 1962, Houbolt's idea stirred up a storm. "Your figures lie!" shouted one NASA engineer. "He's being misleading." Von Braun shook his head at Houbolt and said, "No, that's no good."

Decision for Lunar Orbit

Houbolt persisted. Back at Langley, he got his committee to help him write a two-volume report on LOR, 100 copies of which were distributed around NASA. One early convert was Robert R. Gilruth, chief of the Space Task Group at Langley. But it was Houbolt's letter to Seamans that set NASA to rethinking its landing concepts. "I was not satisfied with the proposals for the lunar landing," Seamans said a few years later, "and I wanted to make sure that we were not overlooking a good bet."

Seamans, therefore, transmitted the letter to Holmes with a request that it be given serious consideration before any final decisions were reached. Holmes passed it on to Joseph Shea, his newly hired deputy who was destined to become the Apollo spacecraft manager. After a visit to Langley in early 1962, Shea was won over to LOR and, within a few weeks, so were Seamans, Holmes and even von Braun. "We swung in favor of LOR mainly because using this method would require only half the weight of the other concepts," Seamans explained later.

In July, Seamans authorized Holmes to initiate the development of systems for the vehicles required for a lunar-orbit rendezvous mission. But no one seemed quite willing to make a categorical statement in favor of LOR.

Pressure for a decision was building up. North American Aviation, which was chosen as the Apollo spacecraft contractor in December, 1961, had found that it could not proceed further with its design and development work until NASA made up its mind on which way the mission would go. North American preferred the direct ascent method; one unspoken reason was that to elect LOR would lead NASA to bring in another major contractor to build the landing craft. The company had an influential ally in Jerome Wiesner, President Kennedy's science adviser and a man who had always been cool to Apollo.* In late July, Wiesner and the President's Science Advisory Committee issued a report questioning the choice of LOR and recommending further study of a direct-flight scheme requiring smaller rockets than the Nova. NASA found itself stymied.

Those were days when it seemed as if Apollo might never rise above the spirited jockeying for position that is perhaps inevitable in the opening stages of any great endeavor. "The decision on how to get to the moon," Houbolt has

---

* Long after he left the post of presidential science adviser, Wiesner generally refused to talk with newspaper reporters about his role in Apollo decisions.

said, "was fraught with a lot of intrigue that has never come out."

The bitterness of the squabble, which involved personalities as well as technical alternatives, burst into the open in September, 1962. President Kennedy was making a tour of the space centers, accompanied by Webb, Wiesner and some foreign dignitaries. At the Marshall Space Flight Center in Huntsville, Alabama, they were being guided through the rocket-building works by the center's director, Wernher von Braun. As von Braun began to brief them on the lunar-orbit rendezvous scheme, Wiesner broke in with a sharp "That's no good." Kennedy stood in embarrassed silence while von Braun, the convert to LOR, and Wiesner, the last influential holdout, engaged in a heated debate.

The controversy prompted John W. Finney of *The New York Times* to write on October 13:

> Some eighteen months after the orders for a lunar expedition were first given by President Kennedy, a spirited interagency argument is still going on within the Administration over the best way to reach the moon.
>
> There are growing indications that the continuing argument is not only causing dissensions in the management of the program . . . but also is threatening delays in the high-priority effort to land a manned expedition on the moon before the end of the decade. . . .
>
> The point being made by Dr. Wiesner is that the earth rendezvous and direct ascent had previously been dismissed by the space agency on the assumption that three men would be landed on the moon.
>
> Now that only a two-man expedition will be sent, it may become more feasible, just from a weight standpoint, to send them by a direct ascent with the advanced Saturn 5 rocket, which will be used for the lunar-orbit trip.
>
> The space agency has conducted some feasibility studies on a direct ascent or earth-orbit rendezvous with two-man capsules.
>
> As far as a direct ascent is concerned, it appears that it would be beyond the capacity of the Saturn 5 rocket unless hard-to-handle hydrogen fuel were used all the way. Present plans call for using storeable fuels, which have less energy, for lowering the capsule on to the lunar surface and launching it from the moon. . . .

Privately, space agency officials are complaining that the Wiesner staff is "meddling" in an area in which it may have some authority but certainly no responsibility.

After the Wiesner-von Braun argument, an exasperated President Kennedy reportedly told James Webb: "Mr. Webb, you're running NASA—you make the decision." Accordingly, in October, 1962, Webb sent a letter to Wiesner indicating NASA's justification for lunar-orbit rendezvous. Holmes summed up the reasoning: "The schedule advantages, cost advantages, and development simplicity of LOR led to its selection as the prime mode." On November 7, NASA publicly announced affirmation of the July decision for LOR and the selection of the Grumman Aircraft Engineering Corporation as the contractor to build the landing craft.

Out of the smoke of battle emerged the first clear outline of how the United States intended to fulfill Kennedy's pledge. The basic components for the flight, following the lunar-orbit rendezvous method, would be a Saturn 5 rocket (a three-stage rocket with a thrust of 7.5 million pounds at lift-off) and a spacecraft with three sections—the Command Module (where the three astronauts would ride), the Service Module (the equipment and rocket unit behind the Command Module) and the Lunar Module (the detachable craft in which two men would ferry to and from the moon's surface).

Having made the decision of how to get to the moon, the Apollo project planners were then confronted with the enormous task of mobilizing the men and resources, designing and building the machines and perfecting the techniques for the mission whose flight rules grew out of the genius of Kepler and Newton and whose hardware had been anticipated in the dreams and experiments of Tsiolkovski, Goddard and Oberth.

Chapter IV

Gathering the Team

Hotspur, the man of action in Shakespeare's *King Henry IV, Part I,* talked a good flight hundreds of years ago:

> By heaven methinks it were an easy leap
> To pluck bright honour from the pale-fac'd moon.

But during the eight years of preparations for Apollo, I never met an engineer or astronaut, construction worker or computer programer, metalworker or project manager, who would agree.

To pluck honor from the moon required of the United States a mobilization of men and resources unprecedented in the annals of exploration. Columbus dared the Atlantic with only three frail wooden ships. Daniel Boone crossed the mountains into Kentucky with five men, their rifles and the provisions they could carry on horseback. Lewis and Clark headed up the Missouri, to reconnoiter the lands of our "manifest destiny," with a keelboat and two smaller craft, knives, tobacco and beads for the Indians—all worth no more than the $2,500 allowed by the government for their expedition. Richard Burton set out in search of the Nile headwaters with a caravan of 170 men and a portable sundial, a pedometer, 2,000 fishing hooks and four umbrellas. Charles Lindbergh, a loner like Boone, needed only his personal savings of $2,000 and $13,000 raised by St. Louis businessmen to take off for Paris and immortality.

Man's first voyage to the moon was a team effort involving directly some 400,000 people and about 20,000 industrial and university contractors. Among them were some of the country's foremost scientists and engineers. It could not have been otherwise. The distant moon would not give up its secrets easily. No single man, no matter how talented and inspired, could have carried off a lunar landing. No group of businessmen in St. Louis or any other city or nation could have raised the necessary $24 billion.

The mobilization started immediately after the Kennedy decision in 1961 and gathered momentum after the selection of the lunar-orbit rendezvous method for reaching the moon. President Kennedy was so anxious for "full steam ahead" that he imposed the highest priority rating, DX, which only a tiny handful of projects, most of them military, had ever carried before.

Abandoned war factories sprang to life as rocket assembly plants. A Massachusetts hosiery mill was turned into a laboratory for guidance instruments. Piny woods in Mississippi were cleared for rocket test firings. A desolate stretch of sand and palmetto scrub in Florida was transformed into a billion-dollar moonport. A network of tracking stations was strung around the world, on ships and islands, at remote sites in Africa and Australia. A pasture outside Houston was bulldozed for the flight control center, a complex of training simulators and test chambers, electronic computers and control consoles.

It was a tremendous national effort, and the costs were staggering, of course. But as James Webb, the NASA administrator who led the effort, put it, "In a very real sense, the road to the moon is paved with brick, steel and concrete here on earth." That is, the money for Apollo was being spent not on the moon, but on earth. It provided challenge and reward for the nation's engineers. It went to enrich large corporations and their shareholders, to build production facilities for the future, to transform sleepy villages into bustling communities.

And there was an aura of pioneering and romance about it all. The men and women in the factories and the forges, in chambers beneath a launching pad's base, on catwalks where rockets were assembled, all were being asked to create new materials and instruments, new methods of measurement and computation, new and highly sophisticated production techniques. They were being called upon to meet standards of reliability never before required.

Unlike the Manhattan Project to develop the atomic bomb in the United States or Soviet space effort, there was no great cloak of secrecy draped over the Apollo project. People in the plants and laboratories knew what role their component would play in the landing mission. Visitors could get in to see the Apollo team at work. However, a visit was not always a simple matter. There were the armed guards at the gate and at each building inside the gate, whether it was an industrial contractor or a NASA facility. Only after a visitor identified himself to the guard's satisfaction, filled out a form and was vouched for by the person to be visited—only then was he given a badge and waved through. To get into the more critical work areas, where a speck of dust was an intolerable intruder, the visitor had to snap on a cloth cap, don a long white smock, and run his shoes (top and bottom) through a motorized scrubber.

Supervision was meticulous. I recall watching a quality-control man in a blue smock keeping books on each part that went into or came out of a spacecraft. Another engineer inspected the workers who went into the cabin and logged the tools they took in with them. Workers were required to check their watches and all the contents of their pockets with him. As a concession to sentiment, workers were allowed to wear their wedding rings, but they had to be taped securely on their fingers.

The Moon Ministry

The task of planning and overseeing this vast mobilization fell to the National Aeronautics and Space Administration, which had been in existence only since October, 1958, as the organization primarily responsible for civilian space exploration. Created as a response to Sputnik, NASA was a union of disparate groups with one thing in common—long experience in such fundamentals of spacecraft as rocketry and propulsion, spacecraft design, tracking and communications. These pools of ready talent formed, in turn, the nucleus of the Apollo project.

The first group embraced by NASA was the old National Advisory Committee for Aeronautics, the so-called "Langley crowd." The committee (NACA) had the longest roots, having been established in 1915 with an initial budget of "$5,000 a year, or so much thereof as may be necessary." Over the decades NACA expanded and revolutionized aircraft-wing design, built and fired rockets for upper-atmosphere research and helped design the X–15 rocket plane, a precursor of manned space vehicles. By the time it was incorporated into NASA, it had grown into an organization with 8,000 employees and five operations and research centers.

At NACA's main laboratory, the Langley Research Center in Hampton, Virginia, worked many engineers who would take leading parts in Apollo—Robert R. Gilruth, Christopher C. Kraft, Jr., Maxime Faget, Caldwell Johnson, Robert Piland, as well as John C. Houbolt, whose prodding led to the adoption of the lunar-orbit rendezvous techniques. Hugh L. Dryden, NACA's director for 11 years, was deputy administrator—the no. 2 man—of NASA during its formative years and the period of Apollo's mobilization.

Next to be drawn under NASA's umbrella was the California Institute of Technology's Jet Propulsion Laboratory, which had been working primarily for the Army. JPL was transferred to NASA in December, 1958, to be owned by the space agency but operated by Caltech on a contract basis. Founded in 1936 by Theodor von Kármán, the brilliant Hungarian-born aerodynamicist, JPL gave American engineers their first organized opportunity to experiment in rocketry. This led, during and after World War II, to JPL's pioneering experiments in radio-guided missiles, techniques for transmitting instrument data via a radio link (telemetry) and the development of prototype earth-orbiting satellites, including the first American spacecraft, Explorer 1. Much of this specialization in guidance and communications reflected the interest of JPL's director, William H. Pickering, a New Zealand-born electrical engineer whose eyes twinkle with excitement when he talks of interplanetary travel.

"The Germans" were the most highly prized talent in the third major group to become part of NASA. They were Wernher von Braun's team of rocket specialists. During World War II they had put together the first workable ballistic rocket, Hitler's V–2, at Peenemünde on the Baltic Sea. After Germany's defeat, the U.S. army brought von Braun and 118 aides, their rocket blueprints and some captured V–2s to Fort Bliss, Texas. The transplanted Germans tested their V–2s out on the desert at White Sands, New Mexico, and shared their knowledge with army engineers. In 1954, von Braun and his group were moved to Huntsville, Alabama, where they began developing Jupiter and Redstone missiles at the Army Ballistic Missile Agency's Redstone Arsenal. It was to the Jupiter that the nation turned to launch its first artificial satellite and to the Redstone for launching the first two manned Mercury capsules. In July, 1960, despite the army's protests, President Eisenhower transferred the "Germans" (who were by then naturalized citizens) and 4,600 engineers and workers to NASA. The Huntsville operation was rechristened the George C. Marshall Space Flight Center and was given primary responsibility for developing the NASA rockets that would provide the muscle for Apollo. (See Chapter X.) NASA also gathered in the Naval Research Laboratory's Vanguard scientific satellite project. The Vanguard team, headed by John Hagen and some other NRL scientists (Homer Newell, John Clark and Robert Jastrow), formed the core for a new NASA center near Washington, the Goddard Space Flight Center, at Greenbelt, Maryland. Goddard was to become the communications center for Apollo's worldwide tracking network.

For all their impressive credentials, these engineers and scientists had little experience managing broad, complex projects. No one, of course, had ever had to come to grips with a project as comprehensive as Apollo. But there was a new breed of engineers, called "systems engineers" and "systems managers," who came closest to understanding how to judge the scientific and engineering validity of other men's concepts and how to schedule the work of thousands of contractors so that all systems meshed into one, and did so on time. More and more, as Apollo gathered steam, NASA called on the services of such engineers.

The language of these engineers was the shorthand of getting things done. They spoke of "trade-offs"—the choosing of systems through a series of compromises dictated by money and schedules, manpower and material capabilities. They spoke of "interfaces"—the points of contact between components or, in a management sense, between contractors. They spoke of the "state of the art"—existing technology—for one of the restrictions on their work was that they were not to wait for any major "breakthroughs." And these men swore by a system of keeping track of everything —a system they called PERT, for Program Evaluation and Review Technique.

PERT was honed into a valuable management tool by the navy during its crash project to develop the Polaris missile for submarines. For PERT to work, it was necessary for project managers to sort out all the tasks that had to be done (40,000 "key events" in the work schedule for the Saturn 5 launching complex, for instance), establish when they had to be done and in what sequence and determine how long each should take. The result was a row of parallel paths charted from the beginning of the project, running past critical points and converging toward the end as the finished and tested product or facility. In this way, through the help of computers, it was possible for a manager to tell at any given time the progress of the project, where delays were occurring, and how a delay would affect interrelated events. In September, 1961, NASA adopted PERT for all of its activities, especially for Apollo. Without PERT there might have been chaos.

Many of the systems engineers who joined the Apollo team had gained their experience while working on various missile and weapons projects during the 1950s. One of the first and most capable of this new breed was Robert C. Seamans, Jr., who had been chief engineer of the Radio Corporation of America's Missile Electronics and Controls Division. Seamans, who had earned advanced engineering degrees at the Massachusetts Institute of Technology, was brought into NASA before Apollo, in late 1960, as associate administrator. As such, he became NASA's general manager, the man who went to industry and college campuses to recruit managerial talent and who laid out many of the guidelines for mobilizing the Apollo team.*

The man who directed the Apollo project in the early years was D. Brainerd Holmes, who gave up a $50,000-a-year RCA job to become the $21,000 director of the Office of Manned Space Flight in September, 1961. Holmes had built his reputation on such military projects as the navy's Talos missile, the Atlas ICBM and the Ballistic Missile Early Warning System (BMEWS). His deputy director was Joseph F. Shea, who joined the Apollo team at the age of 35 after managing the guidance system development for

---

* After Dryden's death in 1965, Seamans was promoted to deputy administrator. He resigned from NASA in early 1968 and, in 1969, became secretary of the air force.

the Titan missile. Shea's main task was to analyze mission choices and the systems and equipment considered for use in the Apollo program. He was an easygoing man, given to wearing bright red socks and uttering outrageous puns, but his studies and analyses laid the engineering foundations upon which Apollo was built.

Making it Work

All this was only the nucleus. New facilities would have to be built. Aerospace corporations, with their experience in aircraft and missiles and their own systems engineers, would have to be brought in for a major share of Apollo's development and production. How could such a conglomerate-type operation be made to work?

The man with the staggering responsibility was James Edwin Webb, NASA's administrator. He was neither a scientist nor an engineer. His credentials were managerial and political, a background that was to stand him in good stead as he struggled with great energy and dedication to fulfill the Kennedy goal.

A blue-eyed stocky man of 5 feet 9 inches, Webb was born in a small North Carolina town with the improbable name of Tally Ho. He was one of five children whose father was a county school superintendent. After graduating from the University of North Carolina, Webb had a year of law study and a tour of duty as a Marine Corps aviator before he made his first appearance in Washington as secretary to a North Carolina congressman. He resumed his studies at night at the George Washington University School of Law, and after getting his law degree in 1936, went to work for Sperry Gyroscope. He rose to vice-president in seven years before being recalled into the marines.

Discharged as a major in 1946, Webb returned to Washington and in four months was named by President Truman as director of the Bureau of the Budget. Truman later appointed Webb as undersecretary of state, a post he held from 1949 to 1952 while he made some influential friends, including Senators Lyndon B. Johnson and Robert S. Kerr. After the Republican victory in 1952, Webb went to Oklahoma City to work for Senator Kerr as director and assistant to the president of Kerr-McGee Oil Industries.

It was no coincidence that President Kennedy, in 1961, turned to the 55-year-old Webb to head NASA. Webb's friend, Lyndon Johnson, was vice-president, and his employer, Senator Kerr, was chairman of the influential Senate space committee. "I didn't want the job," Webb recalled to me later. "I came to Johnson's office and Phil Graham [the late publisher of the *Washington Post*] was there. I said, 'Look, Phil, dammit, how do I get out of this?' And Phil said, 'There's only one man in Washington who can get you out of it, that's Clark Clifford' [close adviser to Presidents Kennedy and Johnson]. So I went to see Clark and he said, 'I'm the one who's been proposing you for the job.' So I went to see President Kennedy. He said he didn't want a scientist but someone who could *manage* the job."

Soon after Webb's appointment, the Apollo project was launched, and the job of managing NASA became even more demanding. For in addition to the moon venture—which took half of NASA's budget—the space agency was tussling with a variety of other pioneering projects such as communications satellites and unmanned probes to the vicinity of Mars and Venus.

Inevitably, Webb had his share of critics. At first, scientists were suspicious of him and somewhat condescending, considering him "only a bureaucrat." Contractors sometimes complained that it was hard to get a straight, clear decision from Webb. But the NASA boss eventually won over the scientists, proved uncanny in his ability to pick excellent men for his staff, and generally impressed Congress with the orderliness of NASA's presentations and by the fervor of his defense of space exploration as a vital force in the nation's evolution.

In 1967, *Fortune* assessed NASA's management of the space program and concluded that "the evidence is persuasive that the nation has been well served" by Webb and his team. President Johnson was said to have remarked at a dinner that "the best administrator I've got is Jim Webb."

Webb stayed on as administrator until October, 1968—through the internal struggles, some organizational shakeups, and periodic congressional criticism—guiding Apollo to within a year of the landing. Shortly before his resignation, I spent more than two hours with Webb in his Washington office. He outlined for me some of the major decisions he made during the first months of Apollo to set the pattern for the mobilization.

"One of the first things I decided," Webb said, "was that I'm not going to act alone." He, therefore, formed a triumvirate in which he was the "outside man," the agency's contact with the White House and Congress; Dryden was the link with the scientific community, and Seamans was the general manager and the main contact with industry. "The three of us wrestled with every decision," Webb said. "Three men are not as likely to make a mistake as one man acting alone."

The trio's initial job was to decide how NASA was going to get the Apollo work done. Webb wanted to avoid the so-called "arsenal" concept of creating manufacturing facilities owned and operated by the government. This was common during World War II and was still practiced by the army at such places as Huntsville. The air force, on the other hand, farmed out nearly all of its work to private industry, giving contractors such a free hand that they often took an important part in the decision-making process involving military strategy. Webb was aware of the criticism this system had aroused in the press and in Congress. Moreover, Webb did not want to be "the prisoner of any one contractor."

NASA's solution, Webb said, was to "use industry to the fullest extent possible." More than 90 percent of the Apollo funds were channeled to industrial contractors to design and build the hardware and, in some cases, help operate some of the NASA facilities. "But we developed enough in-house capability," Webb added, "so that we could evaluate the contractors' work and costs and keep control of the decision-making process."

In the summer of 1961, Webb created a system of evaluation boards to advise the triumvirate on the proposals submitted by contractors bidding for a share of the Apollo work. These boards, composed of NASA engineers, would rate a company's proposal in terms of technical competence, feasibility, cost and management capability. Their recommendations would then go to Webb, Dryden and Seamans for final action. "Not a single major procurement out of all the billions of dollars worth we've done has ever been upset in a court," Webb said with pride in 1968.

The Primes

By the end of 1961, NASA had awarded nearly all of the prime contracts for Apollo hardware. The "primes" began picking their subcontractors, and the subcontractors began placing orders with smaller suppliers. Companies in 47 states became involved. NASA contracted to the Massachusetts Institute of Technology Instrumentation Laboratory the job of planning and building whatever kind of navigation system Apollo would need. North American Aviation, Inc., was selected as prime contractor for the Apollo spacecraft. For the Saturn 5 rocket, which was agreed on as the launch vehicle, pending the decision on how to go to the moon, the Boeing Company was picked to build the first stage, the Douglas Aircraft Company the third stage, and North American the second stage and the engines for all three stages.*

Since the contracts were generally awarded to the lowest bidder, one of the favorite jokes of the project concerned one of the early astronauts. Asked what he had thought about during the wait for lift-off, the astronaut replied: "I just kept looking around at all those dozens of instruments in front of me and reminding myself that every one was supplied by the lowest bidder."

As the industrial team took shape, Webb realized that NASA itself lacked the desired in-house capability to supervise all the new contractors. One of his handicaps was that of attracting enough experienced managers at the relatively low salaries paid by a government agency. In February, 1962, therefore, Webb wrote to the president of the American Telephone & Telegraph Company, explaining NASA's problem and asking the Bell System to assist NASA. AT&T agreed, and organized a separate subsidiary company, Bellcomm, Inc., to do the work. Many of its engineers were drawn from the Bell Telephone Laboratories, where both Holmes and Shea had worked early in their careers.

NASA also contracted with the General Electric Company to assist in the integration of the Apollo hardware,

---

* As a result of mergers during the decade, North American Aviation became North American Rockwell Corporation and Douglas became the McDonnell Douglas Corporation. Grumman Aircraft Engineering Corporation, as was explained in the previous chapter, was not selected to build the Lunar Module until late 1962.

including the design of automated check-out equipment at the launching pad. GE thus had the "interface" assignment.

NASA's Fertile Crescent

Much of Apollo's most visible brick, steel and concrete was implanted along a thousand-mile crescent from Florida's sandy shores to the Texas ranchlands below Houston. The decisions leading to the development of this so-called NASA crescent were reached by the space agency during the summer and fall of 1961.

Even before Kennedy's Apollo announcement, Kurt H. Debus, then director of the Marshall Space Flight Center's launch team, huddled in March, 1961, with several associates at his office on Florida's Cape Canaveral. Debus foresaw the need for expanding the launching facilities there. The existing pads were built to handle nothing larger than the Atlas, with its 388,000 pounds of thrust. Debus, a veteran of the von Braun group of Germans, instructed his engineers to draw up plans for launching facilities to accommodate rockets of between 5 million and 10 million pounds thrust. This was the beginning of Launching Complex 39—the moonport.

Since the first rocket, a combination of a captured V–2 and an army WAC Corporal for the second stage, roared off Cape Canaveral's sands in July, 1950, launching techniques had remained essentially unchanged. They provided for the assembly, preflight test and erection of rockets and spacecraft on the launching pads. These were stable, concrete bases equipped with steel service towers, or gantries, which permitted technicians to work on the vehicle in the vertical position by means of access elevators and work platforms like painters' scaffolding. Debus and his associates decided the old ways were inadequate for the larger, more complex rockets. They had heard that the Russians did most of their assembling and inspecting indoors, because of the more severe Russian weather, and then moved the completed structure to the pad. Perhaps the Americans should do the same. This would reduce the threat of storm damage and the corrosive effects of the salt atmosphere on the rocket. This would also make it possible to haul a fully assembled rocket out to the pad within days after another rocket had been launched, thus reducing the time between launchings—the so-called "turnaround time."

Seamans directed Debus to pursue the studies, but also to get busy finding a site for launching the moon rocket. With Lieutenant General Leighton I. Davis, Debus considered locations in Hawaii, California, Brownsville, Texas, Cumberland Island off Georgia, Mayaguana Island in the Caribbean, Christmas Island in the Pacific, White Sands, New Mexico, and Merritt Island next to Cape Canaveral.

Climate was one consideration; the base would have to be operational 12 months a year. And it would have to be on or near the water, for the largest components of the moon rocket could only be transported from the manufacturer to the launching site by barge. An island launch base, therefore, offered several advantages. There would be no danger of spent rockets falling on populated territory. But most of the islands, it was decided, were too remote from rocket and spacecraft production centers. Acting on a report by Debus and Davis, NASA concluded in August in favor of Merritt Island. It was next door to existing launching facilities and to the chain of tracking stations already spotted on islands downrange from the Cape area. This, NASA reasoned, would avoid some costly duplication of facilities. Beginning in 1962, NASA acquired 87,763 acres on Merritt Island and the rights to use additional acres of submerged lands, all at a cost of about $80 million.

The point of embarkation for man's voyage to the moon lay, along with Cape Canaveral, in the wedge of land that juts out into the Atlantic Ocean about midway down the coast between Jacksonville and Miami. When NASA moved in, Merritt Island was largely uninhabited, a place of orange groves, mosquito-infested lagoons and flocks of wild birds. The brackish Banana River separated the island from the rest of Cape Canaveral, and the Indian River separated both areas from the Florida mainland.

Until the Space Age, the surrounding communities of Cocoa and Cocoa Beach, Titusville and Rockledge existed on oranges and grapefruit, the fish and shrimp from nearby waters and Patrick Air Force Base. Staying out of the mainstream seemed to be a tradition with them. But in the early 1960s they were invaded by young engineers and rocket builders (median age: 26) who swarmed over the

area to build the moonport from elaborate blueprints
taken from rough sketches drawn by Debus. *Architectural
Forum* described this work as "one of the most awesome
construction jobs ever attempted." The principal features
of the complex were to include:

- A hangar big enough to house the Saturn 5 rockets,
  each standing 363 feet tall after the Apollo spacecraft was mounted on top.
- A mobile launch base on which the rockets would
  be assembled inside the hangar and from which they
  would be launched at the pad.
- A method of transporting rockets and launchers—
  together weighing 12 million pounds—a distance of
  3½ miles to the firing site.
- A service structure enabling technicians to complete
  preparations of the spacecraft and the rocket fueling
  at the launch site.
- A control center, with high-speed computers, from
  which all of these operations could be monitored and
  the countdown controlled.

The building of the moonport brought a gold-rush air of
hard work and hard play to communities like Cocoa Beach.
Motels, restaurants and night spots sprang up along the
beach strip, catering to the thousands of construction
workers, engineers, newsmen and contractor officials. "It
may be hard to accept," a lieutenant colonel at Patrick
Air Force Base told a *Times* reporter, "but this loud beach
is going to be a piece of Americana for the next five hundred years."*

When NASA officials looked around for a place to build
the moon booster rocket, they found some ideal hand-me-
downs along the crescent. Located 15 miles northeast of
downtown New Orleans was the enormous Michoud plant,
an army ordnance factory that had nearly 2 million square
feet of manufacturing space. Its one roof covered 43 acres.
The plant had been built by the government during World
War II to assemble airplanes, but was not really used until
the Korean War when the Chrysler Corporation manufactured tank engines there. In 1961, however, it was vacant,

---

* After President Kennedy was assassinated on November 22, 1963, the
name of the Cape was changed from Canaveral to Kennedy. A community
between Cocoa Beach and the launching sites retained the name Cape Canaveral. The Merritt Island facilities were designated John F. Kennedy Space
Center.

a sprawling white elephant. NASA had no trouble con-
vincing the government to permit the Marshall Space
Flight Center to take over the plant and convert it for use
by its prime contractors—Chrysler, for the Saturn 1-B
rocket (the small forerunner to the moon rocket), and
Boeing, for the first stage of the Saturn 5.

After the rockets were assembled, they had to be tested.
In October, 1961, NASA arranged to establish a govern-
ment-owned test facility 35 miles northeast of Michoud,
across Lake Pontchartrain in the piny woods of Mississippi.
It was a sparsely settled area easily reachable from Michoud
by water, rail and road. On the edge, in Slidell, Louisiana,
was another white elephant, a $2 million, brand-new Federal
Aviation Agency computer center being abandoned in favor
of a Houston facility. NASA gratefully took it over.

The 13,427-acre Mississippi Test Facility, as it was
called, developed into a kind of Space Age Venice. It was
laced with 15 miles of 200-foot-wide canals by which the
various rocket stages and engines were brought from
Michoud for test firings and then barged to the Cape
launching pads. The stages built in California, at contractors' plants, were shipped by barge through the Panama
Canal and up the Gulf of Mexico for their test firings.

The most controversial of NASA's early decisions on
facilities was the one to establish the Manned Spacecraft
Center at Houston, on the western end of the crescent.
This was to be the home of the Space Task Group (STG),
the men at Langley who were given the job of managing
the development of the Apollo spacecraft and training the
astronauts.

Even before Apollo, the STG had begun to outgrow its
Langley facilities as the manned Mercury project reached
the flight stage. At first NASA officials considered convert-
ing the Ames Research Center in California to a manned-
flight laboratory. Virginians, naturally, hoped NASA would
merely expand the Langley center, According to *An Administrative History of NASA, 1958–1963,* prepared under
the auspices of the NASA historical staff, a selection team
began in late May, 1961, to inspect prospective sites for
the center, although public announcement of the activity
was not made until after the new NASA appropriations
act, including $60 million for the new center, was passed
in August.

Then the decision came swiftly. On September 1, NASA asked the Army Corps of Engineers to manage the construction of a manned spacecraft center to be built in Houston. On September 19, NASA made the formal announcement. "There was considerable speculation," the NASA history noted, "that the selection of the Houston site was influenced by the fact that a Texan, Lyndon Johnson, was Vice President and chairman of the Space Council and that a Houston congressman, Albert Thomas, was the chairman of the House Independent Offices Appropriations Subcommittee, the subcommittee handling NASA appropriations."

The NASA publication did not pursue the point. Why was it necessary to move the spacecraft planning and training operations from Virginia? Why was it better to have the flight control center at Houston instead of at the Cape where the launch control center would be?

NASA's selection team was said to have appraised a number of sites on ten points: availability of education institutions and other facilities for advanced scientific study, electric power and other utilities, water supply, climate, housing, acreage, proximity to varied industrial enterprises, water transportation, air transportation and local cultural and recreational resources. On this basis Houston—with its booming industry, universities and location near the Gulf of Mexico—may well have rated higher than other possible sites. But this never satisfied the critics who suspected pork-barrel tactics were the overriding influence.

In 1962, many of the Langley engineers began moving to temporary quarters in Houston, while construction of the new center got under way. The site was a 2½-square-mile tract of land 26 miles southeast of downtown Houston and, although part of Houston, separated from the rest of the city by several small towns. It would take $193 million to turn the rough ranchland into the campuslike control center for Apollo.

When the center became fully operational in 1965, the Langley engineers were the nucleus in a staff of 4,500 government employees. Robert R. Gilruth, the head of STG, was the center's director. On the fringes of the center, in new concrete and steel buildings, worked another 4,500 people employed by aerospace firms. Shopping centers, space-related industries and whole new industries sprang up around the center.

Stresses and Strains

By the time these mobilization decisions were made, some stresses and strains were appearing in the headquarters operation in Washington. A dispute developed between Holmes and Webb over the interpretation of the DX priority rating which President Kennedy had given the moon project. In 1963, Holmes wanted a bigger share of the NASA budget allocated to Apollo in order to do his job the fastest way possible. He also sought greater authority over the way the project was run. Webb balked. Being sensitive to the budgetary ways of Congress and wanting to keep Apollo from being the tail that wagged the NASA dog, Webb tried to squelch Holmes's expansive rumblings. As a result, the outspoken Holmes handed in his resignation.

Brought in to take Holmes's place was George E. Mueller, the 45-year-old vice-president in charge of research and development for Space Technology Laboratories and a former engineering professor at Ohio State University. Mueller, too, fit the category of systems engineer and had had experience with the Pioneer space probes.

Another move to strengthen the Apollo management came in early 1964 when Samuel C. Phillips joined the team as the Apollo program director. The tall, lean air-force general, who was trained as an electrical engineer, had managed the development of the Minuteman ICBM, an effort with many similarities to Apollo in that advanced technologies and rigid scheduling were guiding factors. Phillips was to emerge as the key decision maker, prodding companies to get their work done on time, monitoring their costs and performance, giving the go-ahead signal for flights.

The Astronauts

The real celebrities at the Manned Spacecraft Center, the men at the apex of this whole national endeavor, were the "glamour boys" for whom all the preparations were made—the astronauts, the men selected to do the actual flying in space.

At the start of Apollo, there were only seven astronauts —the "original seven"—and they were busy training for the one-man Mercury flights. They were M. Scott Carpenter, L. Gordon Cooper, John H. Glenn, Jr., Virgil I. Grissom, Walter M. Schirra, Jr., Alan B. Shepard, Jr., and Donald K. Slayton. Clearly, more astronauts would be needed. For the astronaut was not just the pilot of the craft, but a valued adviser to those developing it. "In all our manned space programs," Grissom once wrote, "we astronauts have been deeply involved in every step of the development of our spacecraft and their supporting systems. If one of us didn't like the solutions the designers came up with, we said so without hesitation. If we thought we had a better idea, we were free to say so, and often did."

Because the success of the whole moon project could turn on the performances of the men who flew the machines, selection of the astronauts was a painstaking, complex, thorough process. When the process began for the "original seven" in January, 1959, there was widespread public curiosity because of the unprecedented risks the men would face and the exotic nature of the training they would undertake. And when the final selections were announced, Americans everywhere wanted to know all about the men and their families: What sports did they like? What did they eat for breakfast? Which ones had been Boy Scouts? What did their wives think about the risks involved? In effect, a new kind of Space Age folk hero was born.

The selection began with a list of 110 men, culled from armed-services records on the basis of piloting skills, physical health and psychological adaptability. When the selection committee notified the men on the list, it was happy to learn that, though there was a strong element of the unknown involved, there was a high rate of volunteering among the prospects.

Through a battery of written tests, technical interviews, psychiatric interviews and medical history reviews, the list was reduced to 32 candidates. These 32 then were put through extreme mental and physical environmental tests. In the personality and motivation studies, the candidates were asked to answer such questions as "Who am I?" and "Whom would you assign to the mission if you could not go yourself?"

The list was narrowed to 18 candidates for final evaluation by a committee at Langley. The qualifications of these men were so nearly equal in most respects that the ultimate determination was based on their technical abilities and the technical requirements of the program. An important factor in the choice of the finalists was how each would complement the others chosen.

Finally, by mid-April, the selection process was completed and seven men were tapped to be the pioneers in space. Each was telephoned individually and was asked if he still wanted to become a Mercury astronaut. All said yes. As one of them put it, "How could anyone turn down a chance to be a part of something like this?"

The crews that would be most deeply involved in developing and flying the Apollo spacecraft—that is, after the Mercury phase—came after the "original seven" and included scientists as well as pilots and civilians as well as servicemen. In September, 1962, NASA made its first of four expansions of the astronaut corps. Among the nine chosen was Neil A. Armstrong, who was to become Apollo 11's commander. A year later, 14 more astronauts were added to the corps, including Edwin E. Aldrin, Jr., and Michael Collins, Armstrong's mates aboard Apollo 11. Six scientist-astronauts were chosen in 1965, and 19 pilots were selected in 1966.

NASA listed the following qualifications for its astronauts after the Mercury group:

- Selectees must be in perfect physical condition, and are required to take part in a limited space simulation program after a thorough medical and psychological examination.
- They must be no older than 35 at the time of selection, if qualifying for pilot-astronaut, and no older than 36 if qualifying for scientist-astronaut.
- All must be United States citizens and no taller than six feet. Pilots must have a bachelor's degree in engineering, physical or biological science, while scientists are required also to have a doctorate in natural science, medicine or engineering, or the equivalent in experience.

- Pilots must have acquired 1,000 hours' jet time or have been graduated from an armed-forces test-pilot school. Scientist-astronauts not already jet-qualified are required to undergo jet pilot training.

The many and varied pieces that went to make up the Apollo undertaking—the astronauts, the project managers, the engineers, the factory workers, the flight hardware—began to fall into place by 1963. Meanwhile, other engineers and workers were already taking steps to learn what kind of target the moon was and how well men could fly their vehicles in space. It was all part of the gigantic effort to land men on the moon that would cost the United States nearly $24 billion for Apollo and another $1.6 billion for such pioneering projects as Mercury and Gemini—for a total of $25,591,400,000. The following estimate of costs for the manned lunar landing program was prepared by NASA in January, 1969.

APOLLO

Apollo Spacecraft	$7,945,000,000
Saturn I Launch Vehicles	767,100,000
Saturn IB Launch Vehicles	1,131,200,000
Saturn V Launch Vehicles	6,871,100,000
Launch Vehicle Engine Development	854,200,000
Mission Support	1,432,300,000
Tracking and Data Acquisition	664,100,000
Ground Facilities	1,830,300,000
Operation of Installations	2,420,600,000

GEMINI

Spacecraft	797,400,000
Launch Vehicles	409,800,000
Support	76,200,000

MERCURY

Spacecraft	135,300,000
Launch Vehicles	82,900,000
Operations	49,300,000
Tracking Operations and Equipment	71,900,000
Facilities	53,200,000
TOTAL	**$25,591,900,000**

Chapter V

The Pathfinders: Unmanned

The Apollo undertaking was as audacious as it was vast, complex and expensive. Consider how little was known in 1961 about space travel. No American had so much as orbited the earth. No American spacecraft had journeyed to the moon. No one, neither the scientists nor the engineers, could be sure that the moon was a safe place for a landing. There was only the Kennedy decision, the reservoir of resources and native American self-confidence about things technological.

Yet once the Apollo decision was made, Americans reached out and in an incredibly short span of time touched the moon. They photographed its surface, picked at its soil and analyzed its rocks. They circled the desolate body many times, testing its gravitational forces, mapping potential landing spots. At first they didn't go there themselves, of course. They sent mechanical envoys.

One of the initial steps after the Apollo decision was taken was to see what the moon was like, whether or not it was safe to land men there and, if so, where would be the best landing sites. To answer these questions the United States mounted three unmanned lunar exploration projects in the 1960s—Ranger, Surveyor and Lunar Orbiter.

These robot scouting parties for Apollo were controlled from a quiet installation nestled against the San Gabriel Mountains on the outskirts of Pasadena, California. This is the Jet Propulsion Laboratory. One of the nation's oldest space-research centers, it built the first American spacecraft, Explorer 1, and designed vehicles to pass by Venus and Mars as well as to go to the moon. "We have friends in high places," JPL once boasted in a recruiting advertisement.

At a distance, as one approaches JPL from the Rose

Bowl two miles away, there is nothing particularly cosmic about the concrete and steel buildings stair-stepping the slope, pushing back the greasewood and yucca. The installation might be taken for an electronics plant or small engineering college. But inside the Space Flight Operations Facility, a windowless building packed with computers, communications circuitry and control consoles, it is possible to look over the shoulders of intense young men in shirtsleeves as television pictures of the moon flash on the screen and humming machines churn out yard after yard of coded messages from a spacecraft tens of thousands of miles away.

This, in the Space Age sense, is the music of the spheres. To spend much time at JPL is to feel the tempo at which these men work. They are generally young (in their 30s), fiercely independent and not a little cocky. The JPL *esprit* provoked one scientist to comment, "It's almost offensive. It's like the marines." But he did not fault JPL's dedication to space exploration or its capabilities.

Even so, exploring the moon with unmanned craft turned out, for JPL and others, to be no easy task. The United States recorded 12 moonshot failures, with Pioneers and Rangers, before finally succeeding with Ranger 7 in 1964. The Soviet Union was only slightly more successful. One of its vehicles, called Luna 2, smashed almost dead center into the front side of the moon on September 12, 1959, to become the first man-made object to touch another celestial body. Three weeks later Luna 3 flew around the back of the moon, returning the first photographs of the side always hidden from earth viewers. Though rather blurred and grainy, the pictures revealed that the moon's far side, like its front, was pitted with craters. Subsequent Soviet moonflights, however, were plagued with failure until 1966.

Ranger

The Ranger project, the preliminary step in America's lunar exploration, was actually initiated before Apollo. In 1959 NASA officials asked JPL to design a spacecraft capable of returning pictures of the moon's surface from up close. The craft they had in mind would help the nation develop a moonflight capability and also give scientists a clear view of lunar features never before seen by man.

What the JPL engineers came up with was a vehicle that weighed about 800 pounds and, in flight, looked like a giant dragonfly. Ranger's wings were its two solar panels, extending 15 feet from tip to tip. These were designed to absorb solar energy and convert it to electricity to run the spacecraft's instruments. Ranger's ten-foot body tapered from its hexagonal instrument section to a narrow tail housing television cameras. With the new Atlas-Agena rocket as the launcher, the Ranger would first go into an earth orbit and then, with the refiring of the Agena, shoot out toward the moon. After necessary mid-course corrections it would aim at the moon, relaying pictures during the last 20 minutes of flight before crashing into the lunar face.

The first two Rangers, launched in 1961, never got out of earth orbit. The Agena failed to restart. Three more Rangers, more sophisticated models of the first two, were launched in 1962. But they, too, were failures. An aiming error by the Atlas booster caused Ranger 3 to miss the moon by about 23,000 miles. Ranger 4's control system short-circuited, sending the spacecraft crashing into the far side of the moon. Ranger 5's power system went dead before it reached the moon.

This frustrating string of failures roused Congress to investigate the project. How could the nation send men to the moon if it couldn't send robots? As a result, the spacecraft was redesigned to include more backup systems. Ranger 6 followed a perfect course to the moon's Sea of Tranquillity, but when it plunged toward the surface its television cameras were dead. There were no pictures.

After another investigation, coupled with threats of a shake-up of JPL, Ranger 7 was launched on July 28, 1964. It worked. The 807-pound package of six television cameras beamed 4,308 black-and-white pictures back to earth before hitting the Sea of Clouds. They showed craters as small as three feet in diameter and some rocks no more than ten inches wide—features that no earth-based telescope could have picked out. The resolution of the pictures was 700 to 2,000 times better than earth-based photographs of the same area.

"The first Ranger 7 pictures," says JPL's director, William H. Pickering, recalling all the headaches, "were the most emotionally satisfying of all our flight experiences."

In 1965 Rangers 8 and 9 repeated Ranger 7's success, transmitting 12,951 high-quality pictures of the moon. Ranger 8 photographed the Sea of Tranquillity and Ranger 9 focused on the highlands area near the Sea of Clouds.

The Ranger pictures provided the first evidence that many areas of the moon were smooth enough for manned landings. Geologists were surprised, in fact, to find so few boulders and loose rocks on the lunar surface. Still, no one could be sure that the surface was firm enough to support a 15-ton landing craft. From a scientific standpoint the pictures settled none of the arguments over the nature of the moon's surface—whether it was blanketed with dust, was porous and crumbly or was volcanic or meteoritic in origin. "Everyone saw the image of his own theories in the Ranger pictures," recalls John O'Keefe, an astronomer at NASA's Goddard Space Flight Center.

Surveyor

It was left to a spidery, three-legged spacecraft called Surveyor to furnish a more spectacular and conclusive report on lunar conditions. The Surveyor was designed to land softly on the moon, photograph its immediate surroundings, test the "bearing strength," or firmness, of the lunar surface and even run a chemical analysis of the soil.

The job of designing Surveyor fell also to JPL. And like Ranger, it gave its developers years of trouble. Begun in 1960 as a strictly scientific project, Surveyor was nowhere near the flight stage when Apollo planners looked to it for help. Its scheduled 1963 launching date had passed. Project costs were running ten times the original estimates of $72.5 million. Troubles with the spacecraft and the new Centaur rocket forced one delay after another, provoking once again a congressional inquiry. The House Committee on Science and Astronautics criticized JPL, NASA and the prime contractor, Hughes Aircraft, for "poor management."

In their impatience the critics failed to take into full account the difficulty of the task. Even JPL officials concede that they at first underestimated the job. "We had a certain overconfidence in our ability to do things," Pickering once told me. "We didn't give the project enough support in the early days. We didn't recognize the complexity of Surveyor."

Surveyor was the most complex piece of space machinery of its day. Into its triangular aluminum frame were packed fuel tanks, small guidance rockets, various sensing devices and the guidance equipment on which the soft landing depended—radar, computer, autopilot and braking rocket. Its total weight at lift-off was at least 2,200 pounds. After finding its way across 240,000 miles of space, Surveyor was expected, on command from the ground, to point its braking rocket toward the moon at a slant. At about 200 miles above the surface the altitude-marking radar had to switch on and, at 60 miles, the radar had to order the braking rocket to fire. This blast was to slow Surveyor from a speed of 6,000 miles an hour to about 250. Another radar, similar to the one astronauts would later use for their lunar descent, would begin bouncing signals off the moon to help Surveyor feel its way to the surface. It was up to the three smaller rocket engines to apply more braking thrust until the spacecraft was 14 feet off the lunar ground. Then, with all engines shut off, the craft would fall the rest of the way.

While Surveyor was still untried and unproven the Soviet Union was also concentrating on lunar-landing attempts. Its first five soft-landing efforts were failures. Finally, on January 31, 1966, a Soviet spacecraft named Luna 9 reached the moon's surface in working order. The United States would again have to settle for second place. When I called Frank Colella, JPL's public information officer, with the news, he reflected on the car-rental agency that was currently making much in its advertising about being No. 2 to Hertz. "Now I know how Avis feels," Colella sighed.

From Russian accounts it appeared that Luna 9 was relatively crude compared to Surveyor. A sphere little more than twice the diameter of a basketball, it weighed only 220 pounds and had been tossed onto the lunar surface just before its parent vehicle crashed. Bouncing onto the moon, Luna 9's protective covering had dropped away like flower petals exposing a single television camera. The spacecraft returned 27 pictures of rocks and soil before its batteries died.

About two months later, on April 2, Luna 10 became the first vehicle ever to orbit the moon. The 540-pound spacecraft carried instruments to measure the moon's radiation and magnetism, if any, and to determine the density of the moon's crust. But it apparently had no camera, though it did carry a tape recording of the Communist "Internationale," which was beamed back to earth.

An analysis of Luna 10's data produced the first strong evidence that the moon had a crust somewhat like earth's. Luna 10 found that gamma rays typical of basalt were being emitted by the lunar surface. To scientists this indicated that the moon, like earth, was once molten and that its lighter rocks, including basalts, rose to form a crust. The basalts on earth are dark gray, dense, fine-grained rocks.

With high hopes but modest expectations, American space officials gathered at Cape Kennedy in May, 1966, for the launching of the first Surveyor. JPL engineers were particularly tense, for they felt the prestige of their laboratory was on the line. Their fears proved groundless. On June 2, after a flight of more than 63 hours, Surveyor 1 landed softly on the Ocean of Storms near the crater Flamsteed. Its rotating television camera went into action and transmitted 11,150 black-and-white pictures before the long lunar night set in.

Robert J. Parks, the cigar-smoking project manager of Surveyor at JPL, could relax for the first time in five years. So could Apollo planners. From the way Surveyor landed, concluded Parks, "you certainly would expect the Lunar Module to be able to land."

At a news conference in Washington after Surveyor's picture-taking days were over, Leonard Jaffe, JPL's project scientist, said: "The moon surface looks like a soil, not very hard, with rocks and clods on it and in it." Jaffe also observed that the pictures showed clumps of material like wet soil. Such clumping was expected because small particles tend to stick together in a deep vacuum like that on the moon. Other scientists said they identified some pock-

marked rock reminiscent of lava in which bubbles of gas
have burst during cooling.

Four of the remaining six Surveyors were equally
successful. They confirmed once and for all that the
moon was a safe place for manned landings.

Surveyor 2 was launched in September, four months
after the initial success and a month behind another
Soviet moon-orbiting craft, Luna 11. This Surveyor was
aimed at the Central Bay (Sinus Medii), but a failure of
one of its maneuvering rockets sent it into a wild tumbling
flight to a crash landing.

Then, on April 19, 1967, Surveyor 3 landed gently on
the Ocean of Storms and did more than take pictures. It
stretched out a robot arm, flexed its mechanical muscles
and dug a hole in the surface of the moon with a sharp
metal claw. With slow, sometimes jerky movements, the
mechanical arm extended itself nearly five feet on commands radioed from the JPL control room. The experiment, developed by Ronald Scott of Caltech, demonstrated
that lunar soil was fine-grained stuff that tended to stick
together like damp clods. It would present no problems
to astronauts, Scott concluded.

Surveyor 4 was a failure. Signals terminated abruptly
within seven miles of the moon's Central Bay. But Surveyor 5 touched down on the Sea of Tranquillity on
September 10 and another ingenious experiment was
performed.

Surveyor 5 carried a small gold-plated box, which was
lowered to the lunar surface by cable. In the box was
a radioactive material, curium 242. Radiation particles
emitted by this material escaped through a hole in the
box to bombard the surface. Since each chemical element
in the soil reflected the particles in a distinctive way, it was
possible for sensors in the box to identify many of the
soil's constituents.

From the data radioed back to JPL, Anthony Turkevich,
a University of Chicago nuclear chemist and designer of
the experiment, concluded that the moon where Surveyor
sat was geologically similar to earth. Oxygen (in combination with other elements, not as a gas) and silicon were
the most abundant elements. These also are the earth's
most abundant surface elements. And analysis of the data

indicated to Turkevich that the lunar rocks seemed to be like basalt, not unlike the rock common to the Palisades of the Hudson River, the Hawaiian Islands and Iceland.

This was by far the most interesting of Surveyor's findings. It made scientists all the more eager to get their hands on the samples of lunar soil the Apollo astronauts would bring back. It even made a few scientists pause and wonder if perhaps George Darwin had not been right when he theorized that the moon was a chunk of our planet that spun off from the Pacific basin when the earth was young.

Surveyor 6, which landed in the Central Bay on November 9, sent back more pictures and soil analysis. The last of the Surveyors, Surveyor 7, was aimed at the rugged highland region about Tycho Brahe, the relatively new and brightly rayed crater at the south of the moon's face. Not one of the likely astronaut landing sites, it was primarily a target of scientific interest. After Surveyor 7 landed on January 9, 1968, its radiation-scattering device detected less iron there than in the Sea of Tranquillity, but otherwise the moon seemed to be fairly homogeneous.

Thus the Surveyors, after a slow start, proved to be excellent pathfinders for Apollo. Benjamin Milwitzky, the Surveyor program manager at NASA headquarters, said the awkward-looking vehicles "established that the moon is entirely suitable for Apollo landings. We've established that a spacecraft on the moon will not sink in deeply, that an astronaut will not need snowshoes to stay on top."

Lunar Orbiter

At about the same time another series of unmanned spacecraft—the Lunar Orbiters—were taking a look at the moon from a different perspective. From an orbit about 26 to 28 miles above the surface they focused their cameras on a 3,000-mile band along the lunar equator on the side visible to earth. This was Apollo's "area of interest," the zone where astronauts would land, and Apollo planners wanted it completely mapped with detailed photography.

All of the five Lunar Orbiters were successful. "5 for 5" read the self-congratulatory signs in the corridors of the Boeing Company's Seattle plant. Boeing was the prime contractor, working under the direction of NASA's Langley Research Center. The missions were controlled through JPL facilities.

Compared to Surveyor, Lunar Orbiter was a fairly simple spacecraft. In flight it looked a little like a four-leaf clover. The center section was the body, housing the camera and film-processing laboratory, communications equipment and the rocket that fired the vehicle into lunar orbit. Extending from the body were two antennas and four solar panels—the four cloverleaves. The 850-pound craft was launched by the Atlas-Agena rocket.

The photographic system, built by the Eastman Kodak Company, incorporated a wide-angle lens and a telephoto lens. The camera's shutters operated simultaneously so that each exposure produced two pictures—one a medium-resolution shot, the other high resolution. In a medium-resolution picture objects eight yards square, about the size of a boxing ring, showed up clearly. The high-resolution lens picked out details as small as the top of a card table. The pictures were developed on board, converted to electrical signals and then radioed to tracking stations on earth. Each spacecraft was capable of taking roughly 400 pictures.

The first of the Lunar Orbiters was launched from Cape Kennedy on August 10, 1966, and swung into orbit of the moon four days later. Succeeding Lunar Orbiters were sent up at intervals of approximately three months.

On the first three missions, the vehicles circled the moon at the equator, following orbits ranging from an altitude of about 1,150 miles on the far side to about 26 miles on the front side. They were thus able to get their highest-quality close-ups of the front side, where astronauts would be landing.

Lunar Orbiter 1 focused primarily on nine promising Apollo landing sites to the south of the equator. After studying its pictures, scientists reported that the moon "is not a cold, dead, lifeless place" but a still-evolving body "very nearly as dynamic as the earth." Harold Masursky, chief of astrogeological studies for the United States Geological Survey, said: "There are many processes going on—the faulting, or uplifting of material, the weathering by mass wasting, the distribution of material by meteoritic impacts and the construction of features by volcanic processes."

When Lunar Orbiter 2 began circling the moon November 10, 1966, it made a similar survey of sites in the northern portion of the equatorial band. On February 8, 1967, the third Orbiter began reexamining the most promising Apollo sites, taking overlapping pictures to get a three-dimensional effect that would bring out the contours of the surface.

Since the first three Orbiters were so successful, giving Apollo planners all they needed in the way of landing-site pictures, the last two flights were devoted to targets of great scientific interest. Both Orbiter 4 and Orbiter 5 circled the moon at its poles, rather than at its equator, and thus were able to photograph 99 percent of the surface as it rotated under them. Orbiter 4 reached the moon in May of 1967; Orbiter 5 completed the series in August of that year.

Altogether the Lunar Orbiters took a total of 1,950 black-and-white pictures. In addition, they indicated that the moon has no belts of trapped radiation, such as earth's Van Allen belts, and that the micrometeoroid hazard is not significant.

One of the project's most intriguing revelations came not from the pictures but from the tracking data. The spacecraft, it seems, had a curious way of dipping and wiggling when their orbits took them closest to the moon's surface. They were occasionaly drawn three to six miles nearer the surface than they should have been. In 1968, after months of analysis, two JPL mathematicians—Paul M. Muller and William L. Sjogren—offered a plausible reason when they observed that the dips had occurred during passes over circular craterlike seas—the Seas of Rains, Serenity, Crises, Nectar and Moisture.

They reported that this behavior could be explained by assuming that large, dense concentrations of mass lie beneath these seas. They called the concentrations "mascons." According to one theory, the mascons could be the remains of meteorites that melted on impact, cooled and then settled into the moon's surface. Whatever the scientific explanation, it raised the eyebrows of Apollo planners. What effect, they wanted to know, would the mascons have on the astronauts' ability to navigate while orbiting the moon?

By the end of 1967 NASA had narrowed to five the number of promising sites for the first astronaut landing. Two were in the Sea of Tranquillity, on the eastern side of the moon, one in the Central Bay, and two more on the western side, on the Ocean of Storms. As the unmanned pathfinders had discovered, these were safe, firm and smooth areas. Other important pathfinders were to prove that men could master the techniques for getting to them.

Chapter VI

The Pathfinders: Manned

There has been little in the whole of evolution to prepare man to live in space. Beyond our sheltering atmosphere lies a hostile void, airless, laced with radiation and hurtling particles. Men in space encounter the glare of undiluted sunlight and a black frigidity. They must take their own atmosphere with them. They must learn to live in what amounts to a state of free fall, the sensation of weightlessness.

Before the first men went into space there were scientists who expressed doubt that the human system could withstand an experience so foreign to the conditions of life on earth. They foresaw dire consequences. Some said the astronauts might be disoriented by the weird optical effects of space or sent into hallucinations by the silence, the solitude and the absence of familiar sensations. Others said the human skeleton and heart might suffer from the lack of stress that prolonged weightlessness involved. The heart might be weakened by having too little work to do and might not be able to react properly to the strain of coming back to earth. There were also those who feared that the bones would begin to lose calcium, as happens when a person lies in bed for weeks. And—who could tell?—there might be subtler types of damages of a nature no one could predict.

However, a larger number of scientists felt there would be no debilitating effects serious enough to endanger space explorers—at least in flights of reasonably brief duration.

But what about the mechanics of getting men into space? Could engineers anticipate all the things that could go wrong and build spacecraft safe enough for human passengers? No one could be sure in 1961. Every so often a rocket had a terrifying way of blowing up in flames on the launching pad. Spacecraft got into orbit and then, when some switch or component failed, went dead. There was the question of constructing a vehicle light enough to get into space but substantial enough to withstand radiation, the impact of micrometeorites and the heat and vibrations of the return to earth.

The only way to be sure that Americans could build and fly space vehicles designed to carry human beings was to do it. This was the primary purpose of the two manned flight projects that preceded Apollo—Mercury and Gemini.

Mercury

"We are behind," President Kennedy said after the Soviet Union's Yuri Gagarin orbited the earth in April, 1961, in Vostok 1, "and it will be some time before we catch up."

It took about ten months.

On May 5, 1961, three weeks before Kennedy's Apollo speech to Congress, Alan B. Shepard, Jr., became the first American to venture into space—but it was far short of an orbital flight. He rode a Mercury spacecraft 302 miles downrange from the Florida launching pad. During his 15 minutes aloft he reached a peak altitude of 115 miles. Both the astronaut and the capsule were recovered in the ocean, proving that both could survive a short ballistic flight. Shepard was in a weightless state only about four minutes. "The only complaint I have," said Shepard, "was the flight was not long enough."

Less than three months later, on July 21, 1961, Virgil I. "Gus" Grissom piloted another Mercury capsule along the same suborbital path. His 16-minute journey proved that Shepard's performance had not been a lucky fluke. When Grissom splashed down, the explosive mechanism on the escape hatch activated accidentally, blowing the hatch open. Grissom swam free and was rescued by a hovering helicopter but the capsule sank.

Both Shepard and Grissom had been launched by Redstone rockets, which were not powerful enough to boost them into orbit. The Atlas, a proven intercontinental ballistic missile, was still undergoing test firings with Mercury capsules holding chimpanzees and robots. Until the Atlas was "man-rated," no Americans could become true orbiting spacemen.

The Soviet Union put its second manned spaceship into orbit two weeks after Grissom's flight. Gherman S. Titov, a 26-year-old air-force major, circled the earth 17 times at altitudes from 110 to 160 miles and returned safely touching down on land inside the Soviet Union, which was the practice of the Russian cosmonauts. Titov's flight raised some apprehensions about man's ability to adapt to the new environment. Of his experience Titov wrote later: "For the life of me, I could not determine where I was. I felt suddenly as though I were turning a somersault and then flying with my legs up! I was completely confused, unable to define where was earth or the stars. . . . My sense of orientation vanished abruptly and completely."

John H. Glenn, Jr., was the first American astronaut to make an orbital flight. After weeks of agonizing and frustrating delay, he finally walked out to the launching pad before dawn on February 20, 1962. Thousands watched the Atlas and Mercury from the Canaveral beaches. Millions more throughout the country sat before television sets. At 9:17 A.M. the Atlas gave a staccato roar and slowly rose.

Unlike Titov, Glenn found that the feeling of weightlessness was very "pleasant—you could become an addict." At no time did he experience any ill effects. He had no difficulty taking food from squeeze tubes. And the view was "tremendous."

The flight was not without tense moments. During his three orbits trouble developed with the small rockets that automatically control the capsule's altitude. Glenn took over manual control of the rockets. In the control room a warning light indicated that the capsule's plastic-glass heat shield had become unlatched. If the shield fell off prematurely, Glenn would be incinerated when the capsule plunged back into the earth's atmosphere.

A decision was made not to jettison the housing for the braking rockets after they were fired. The hope was that the straps attached to the retro-pack, which passed over the heat shield, might hold the shield in place. When the Mercury—Glenn called it Friendship 7—reentered the atmosphere, the retro-pack flew off in "big flaming chunks." Watching them sail past the capsule window, Glenn felt some "cautious apprehension" that the heat shield might be disintegrating. "Boy, that was a real fireball!" he reported to Mercury control. But the shield remained in place, and Glenn's capsule hit the water near the recovery ship. After three orbits of earth Glenn reported, "My condition is excellent."

The nation welcomed Glenn back with the adulation accorded only the rarest of national heroes. His cool performance, modesty, simplicity and unabashed patriotism had caught the national imagination. There were parades and speeches and gold medals. Newborn infants were named for Glenn in dozens of cities and one unfortunate boy in Utah was said to have been christened "Orbit."

There was also a great sense of relief. The United States had finally put a man in orbital flight. To Apollo planners it was reassuring to know that the first steps at least seemed to be in the right direction.

Several people had an important hand in designing and building the Mercury spacecraft. One of the first Americans to think seriously about manned space-vehicle design was H. Julian Allen, a senior engineer with the old National Advisory Council for Aeronautics, NASA's predecessor. In the mid-1950s he concluded that a long, needle-nosed shape was the worst possible configuration. In those days engineers were wedded to the idea that anything that flew had to be streamlined, yet Allen pointed out that an unstreamlined spacecraft would help solve the problem of bringing the vehicle back to earth from an orbital speed of roughly 17,500 miles an hour. By giving it a blunt end and pointing that end in the direction of flight, it would be possible to slow the returning craft by utilizing atmospheric drag. Moreover, the shock wave formed by the blunt end would help scatter the tremendous reentry heat into the surrounding air.

Maxime A. Faget, a diminutive engineer at the Langley Research Center in Virginia, incorporated many of Allen's ideas and drew up what became the preliminary blueprint for Mercury. NASA's official history of the project gives Faget the principal credit for designing the spacecraft and terms John F. Yardley of the McDonnell Aircraft Company its chief developer. McDonnell was awarded the Mercury contract in January, 1959, and Robert R. Gilruth of Langley was made project manager for NASA.

The main body of Mercury was shaped like a bell. It was 9 feet 7 inches long from the base of the blunt heat shield to the tip of the barrel-shaped nose. It was 6 feet 2 inches wide at its widest point, the blunt end, and weighed about 3,000 pounds. The nose contained the antennas and control systems as well as the parachutes for landing. The blunt end contained the three retro-rockets needed to brake the spacecraft for the descent to earth. The inner shell was made of titanium, lightweight but strong, while the outer shinglelike covering was of an alloy called René metal. To save weight the engineers decided that the cabin atmosphere should be pure oxygen at about one-third the pressure at sea level instead of a mixture of nitrogen and oxygen, which is more like normal atmosphere. Since all military experience in life support at high altitudes had been based on pure oxygen, it was an understandable decision. But it turned out to be fateful.

After Glenn's flight three more astronauts took their turns orbiting in the Mercury. On May 24, 1962, M. Scott Carpenter became the second American to circle the earth. His three-orbit trip, scheduled essentially as a rerun of Glenn's, ended in throat-catching suspense. As a result of a 25-degree error in the capsule's orientation during reentry, Carpenter's vehicle overshot its intended landing point by 250 miles and for a time was out of radio contact with the recovery ship. But the astronaut was safe and the mission a success.

On October 3, 1962, Walter M. Schirra, Jr., took the next ride in Mercury. It lasted nine hours as he circled the earth six times in the most flawless mission to date.

The project was brought to a successful conclusion in May, 1963, with the 22-orbit, 34-hour flight of L. Gordon Cooper, who at 36 was the youngest of the "original seven" astronauts. (Of the first group only Donald K. Slayton, because of a heart irregularity, never got a chance to journey into space.) Toward the end of Cooper's flight, malfunctions cropped up in the autopilot, forcing the astronaut to take over the controls himself. He manually guided the craft all the way down to a bull's-eye splashdown. "Right on the old gazoo," Cooper radioed.

Despite some lobbying by astronauts for an additional and longer Mercury mission, Cooper's was the last of the Mercury flights. After six manned voyages, lasting a total of 51 hours and 40 minutes, Mercury had served its purpose. Men had gone into space and survived. Moreover, they seemed able to work and think while in orbit and there had been no undesirable reactions. The nation had spent $384.1 million and thousands of engineers and workers had devoted years of effort to demonstrate this spaceflight capability.

The Mercury project had scarcely ended before the Soviet Union launched yet another Vostok, its fifth. After the Gagarin and Titov missions in 1961, the Russians had outdistanced Mercury with a four-day manned flight in August, 1962. Andrian G. Nikolayev rode Vostok 3 for 64 orbits. While Nikolayev was aloft, Pavel R. Popovich was launched in Vostok 4 for a three-day, 48-orbit flight. The launching was timed and aimed so that Vostok 4 came within 3.1 miles of Vostok 3. For the fifth Vostok mission, nearly a year later in June, 1963, Valery F. Bykovsky orbited the earth 81 times and was joined in space by the first woman cosmonaut, Valentina V. Tereshkova, in Vostok 6. Though they did not accomplish a rendezvous, the two cosmonauts landed safely 2½ hours and 50 miles apart.

Consequently it seemed that, despite the success of Mercury, the Russians were maintaining their commanding lead in manned spaceflight. But work on the next American manned project—Gemini—was already well under way. It would prove decisive.

Gemini

The birth of the Gemini project went almost unnoticed. It came on December 7, 1961, while attention was riveted on the preparations for Glenn's orbital flight. As conceived by Gilruth and one of his chief engineers, James A. Chamberlain, Gemini was to be a two-man spacecraft in which astronauts would demonstrate that they could not only survive in space but also steer a vehicle through many of the complex maneuvers necessary for a flight to the moon.

According to NASA, Gemini's primary objectives were:

1. To provide a logical follow-up program to Mercury with a minimum of time.

2. To subject two men and supporting equipment to long-duration flight.

3. To effect rendezvous and docking (linkup) between a Gemini and another orbiting vehicle.

4. To experiment with astronauts leaving the spacecraft while in orbit and to determine their ability to work and maneuver during these "space walks."

5. To perfect methods of reentry from orbit and of landing at preselected target areas.

For this purpose the McDonnell Company was called upon to build a spacecraft that was larger and more maneuverable than the Mercury. Gemini, however, bore a superficial resemblance to its predecessor. The crew capsule was shaped a little like a bell and it had a blunt end shielded against fiery reentry. But beneath the outer cover was a far more complex piece of machinery.

The crew capsule, called the Reentry Module, was 11 feet high and 7½ feet wide at the base. The combined weight of the capsule and its Adapter Module, the rear equipment unit, was more than 7,000 pounds. The Adapter Module contained the rockets used to begin reentry, fuel tanks and a fuel cell that replaced short-lived batteries as the main source of electricity. This rear unit was jettisoned just before reentry. The crew capsule contained radar for rendezvous maneuvers and a computer the size of a hatbox for in-flight calculations and automatic control, if desired, of such key operations as reentry.

The Gemini design also saw the introduction of the modular concept of spacecraft construction, a concept which was to save much time and prevent delays during the Apollo project. In Mercury most of the systems were stacked like a layer cake inside the pilot's cabin, and when trouble developed it usually was necessary to disturb a number of components to get at the malfunctioning system. In Gemini major components, such as the fuel cell or the air-conditioning, were arranged in self-contained module packages. If one was found to be faulty, it could be reached by removing an external panel and replaced quickly without affecting other systems already checked out.

The first two Gemini flights were unmanned tests of both the rocket, the more powerful Titan 2 military missile, and the spacecraft. No attempt was made to return and recover Gemini 1, launched into orbit on April 8, 1964. The second unmanned flight, in January, 1965, demonstrated successfully the vehicle's ability to withstand the heat and stress of reentry.

All was now ready for men to go aloft in Gemini and the pressure was on. For Gemini was about a year behind schedule and the Russians had again stolen a march on the Americans with the world's first multi-man spaceflight—three men in Voskhod 1, which circled the earth 16 times in October, 1964. And on March 18, 1965, before Gemini 3 could be launched, the Russians sent up Voskhod 2 for a 17-orbit flight. It had two men aboard and one of them, Aleksei A. Leonov, became the first human being to step outside an orbiting vehicle for a floating "walk" in space.

Less than a week later, on March 23, Grissom and John W. Young, a member of the second "class" of astronauts, selected in 1962, were launched in Gemini 3. Though it was a short flight—three orbits—the two men had time to demonstrate that Gemini was up to its advance billing as a real flying machine. It was capable, they proved, of moving up and down, right and left, backward and forward and even of changing orbits. A set of 16 thruster rockets ringing the vehicle made this possible. Thus Gemini 3 became the first spacecraft, American or Russian, to change the plane and size of its orbit.

With the knowledge that Gemini was flightworthy, NASA pushed ahead with nine more two-man missions in the next 18 months, an average of one every two months.

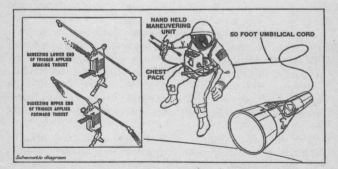

A hand-maneuvering unit, a chest pack and an umbilical cord to the spaceship enabled Gemini astronauts to take space "walks."

On June 3 Gemini 4 was launched on a four-day, 62-orbit flight during which Edward H. White became the first American to "walk" in space. White was so exhilarated by the experience that it took a little coaxing by his commander, James A. McDivitt, to get him back inside the capsule. For 20 minutes he chatted, joked and snapped pictures as he darted about in raw space with the aid of a gas-firing jet gun straight out of Buck Rogers. He caressed the gleaming white titanium hull of Gemini and even wiped the windshield. "This is fun," White exclaimed. Upon returning reluctantly to the spacecraft, White sighed, "It's the saddest moment of my life."

Gemini 4 was the first mission controlled from NASA's new Manned Spacecraft Center at the edge of Houston. The demands of Gemini and Apollo had outgrown the limited facilities of the old control center in a Cape Kennedy blockhouse. The operational heart of the Houston facility is the Mission Control Building, a $7 million structure crammed with $100 million worth of electronic equipment, 10,000 miles of wires and 2 million cross-connections. And the master of the control console in those days was Christopher Columbus Kraft, Jr., who like Gilruth and many others was a Langley alumnus. A Virginian with soft traces of a Tidewater accent, Kraft was then only 41 years old, but already a veteran of every manned mission. He developed almost from scratch the practical arts of directing spaceflights. "We got plenty nervous during those first few launches," Kraft recalls, "because we didn't know enough rules. All we knew was that missiles sometimes came off the pad and went ape."

To be ready for any contingency, all of the "what ifs," Kraft supervised the writing of a rules book, a fat compendium of what to do in the face of all the possible malfunctions during flight. With it as the Bible, Kraft ran his control staff of approximately 600 men through hours of skull practice, called mission simulations or "sims." He made it as hard as possible. Unknown to the men in the control room, even to Kraft himself, computers were programed to introduce artificial emergencies—an oxygen leak, a fuel-cell failure, a radio blackout and so on. They were emergencies that could happen and Kraft wanted his men to know how to react swiftly and calmly. After each simulation Kraft would run a critique in which he did not hesitate to tear apart an associate's performance.

The real test was performance during an actual flight, and with each Gemini the men on the ground as well as the men and machines in orbit demonstrated that they were up to their difficult tasks. Such operational experience became one of the most prized legacies passed from Gemini to Apollo.

Flight controllers learned, for example, how to nurse a crippled spaceship through a successful flight. This happened with Gemini 5. Cooper, the Mercury veteran, and Charles P. Conrad, Jr., a rookie, began their planned eight-day mission on August 21, 1965. Their primary goals were to see if eight days in space (roughly equivalent to the time of a lunar round trip) adversely affected the human body, to test the rendezvous radar system and to check out operation of the electricity-producing fuel cell.

As yet untried in orbit, the fuel cell, producing electricity through the chemical reaction of liquid oxygen and liquid hydrogen, was installed in Gemini 5 because it was smaller and lighter than the conventional batteries used in previous flights. A similar system was already being planned for Apollo.

Early in the first day of Gemini 5's flight, pressure in the fuel tanks for the cell began falling. The cell could not generate its normal output of electricity, leaving Kraft seriously to consider terminating the flight early. But he did not. "If you don't know what to do, don't do anything," he philosophized. Cooper and Conrad and Gemini 5 outlasted the faltering fuel-cell system to complete an otherwise successful mission. Dr. Charles A. Berry, the flight surgeon, reported that the astronauts had withstood their eight-day trip in perfect health.

From the dual mission of Gemini 6 and 7 in December emerged the first strong feeling that man was learning to master spaceflight. It was the turning point of the project.

A failure set up the situation. On October 25, Gemini 6 had been scheduled to attempt the first rendezvous and docking, with an Agena rocket stage as the unmanned target vehicle. But as the astronauts, Schirra and Thomas P. Stafford, waited on the launching pad, the Agena blew up just before reaching orbit. Lacking a target, the astronauts climbed out of Gemini 6 to wait another chance, which many feared might not come for five or six months.

Within five minutes, however, Walter F. Burke, Mc-Donnell's vice-president, was huddling with his top engineers in the Cape Kennedy launch control room and he came up with a plan that could mean a quick recovery from the setback. He proposed to NASA officials the doubling up of Gemini 6 and 7. Gemini 7, slated to be a 14-day vehicle, would be ready by December, and Gemini 6, which was already checked out, could be stored until then. No docking would be possible since only the Agenas were equipped with attachments for linkup. But the first rendezvous could be achieved.

NASA accepted the idea and it worked. On December 4 Frank Borman and James A. Lovell, Jr., were launched in Gemini 7. On December 12 Gemini 6's attempt to leave earth was frustrated by a loose electrical plug in the Titan 2 rocket. The rocket engines started, then shut down with Schirra and Stafford perched atop the Titan. They came within a split second of being the first astronauts forced to eject themselves from a spacecraft during launching.

But three days later, on December 15, all went well and Gemini 6 was on its way to meet the waiting Gemini 7. Schirra steered his ship on a 100,000-mile chase in space, closing in on Gemini 7 at an altitude of 185 miles and a speed of more than 17,000 miles an hour. The two crafts came within a foot of each other and for several hours circled the earth in formation. "A piece of cake," Schirra commented afterward when describing the intricate rendezvous maneuvers.

At the control room in Houston a ripple of relief gave way to unbridled joy. Said Chris Kraft: "It's the biggest milestone since the flight of John Glenn."

Gemini 6 peeled off from Gemini 7 and returned to earth the next day. Borman and Lovell brought their mission, the longest ever flown, to an end on December 18. Doctors could find no serious health problems resulting from the 14 days in orbit.

Even a close shave on March 16, 1966, did not daunt the Gemini team. On that date Gemini 8 was launched and accomplished the first docking with another orbiting vehicle, a 27-foot-long Agena. "Like parking a car," Neil A. Armstrong, the command pilot, said.

Then trouble struck. Gemini 8 began bucking and spinning wildly because of a misfiring thruster rocket. Armstrong and David R. Scott, the other crew member, were forced to make an emergency splashdown in the Pacific before the end of their first day in space.

In the last four Gemini flights, astronauts practiced over and over the techniques of rendezvous and docking and gained more experience operating outside their orbiting craft.

Gemini 9 was the three-day mission in which Stafford and Eugene A. Cernan used three different approaches to rendezvous with a target. No docking was attempted because the protective clamshell shroud failed to break free of the target's docking collar. With jaws wide open, it looked to Stafford like "an angry alligator." Cernan performed a long space walk before the flight ended June 6.

Young and Michael Collins took a three-day trip in Gemini 10, beginning July 18. They achieved the first dual rendezvous, first with their Agena and then with the still-orbiting Agena left over from the Gemini 8 mission. They docked with their own Agena.

On September 12 Gemini 11 was launched. During its three-day flight Conrad and Richard F. Gordon, Jr., linked up with an Agena and then fired the Agena's main engine to propel themselves 850 miles out from earth, the farthest man had ventured at the time.

For the last Gemini flight, Lovell and Edwin E. Aldrin, Jr., raced into orbit November 11 and despite a balky radar unit steered their craft to a successful rendezvous and docking with an Agena. When they splashed down four days later, after numerous rendezvous maneuvers and Aldrin's 5½ hours of space walking, they brought to a successful end the $1.35 billion program.

At a news conference after the splashdown Gilruth declared: "We have done all the things we had to do as a prelude to Apollo."

In the ten Gemini flights the United States amassed a total of 969 hours and 56 minutes in space. The experience in rendezvous and docking was essential to lunar flight planning, which involved such a linkup between the Lunar Module and the Apollo mother ship before the return to earth. Long-duration flight was necessary to prove that man could survive in space for the seven or eight days it should take for a lunar round trip. The extravehicular activity—or space walks—gave astronauts a feeling of what it should be like to work in the low-gravity conditions that exist on the moon and was practice for an emergency procedure should their Lunar Module and Apollo fail to dock properly.

The optimists had been vindicated. "Our outlook to the future is extremely optimistic and man has shown his capability to fulfill a role as a vital, functional part of the spacecraft as he explores the universe," concluded Dr. Berry in a report evaluating the medical results of Mercury and Gemini.

In the postsplashdown enthusiasm Chris Kraft said, "We now have a real confidence that we will be able to compute and carry out the maneuvers necessary to get to and from the moon."

But the self-confidence turned out to be all too premature.

Chapter VII

Apollo's Darkest Hour

When John Glenn returned from his orbital journey he cautioned his countrymen: "We are going to have failures. There are going to be sacrifices made in the program; we have been lucky so far."

Every astronaut had learned to live with the haunting possibility of accident or even death. Every engineer had agonized over the thought that a rocket might explode upon ignition, that a spacecraft might disintegrate during reentry or that a malfunctioning capsule might be stranded in space while the oxygen supply ebbed and the crew died the slow death of suffocation. Posters at the contractor's plant in California reminded workers that astronauts' lives depended on their "PRIDE—Personal Responsibility In Daily Effort." But when catastrophe visited Apollo, it struck in a horrible form that no one seemed to have anticipated. The tragedy came not in space or on the moon but on the ground. It was a shattering blow from which the Apollo project was not to recover for nearly two years.

When it happened, on January 27, 1967, preparations were in full swing for the first flight of an Apollo that would carry astronauts. After months of delay and developmental problems, a launching date had been fixed: February 21, 1967. The three crew members—Gus Grissom, Ed White and Roger Chaffee—had been training for months for the planned earth-orbital mission. The spacecraft had been installed on top of a Saturn 1-B rocket at Launching Pad 34. The time had come for the final check-outs.

The vehicle in which the three expected to accomplish their mission was not, in one sense, the finished product. It had been built according to specifications laid down in

1961, and where engineers came along with better ideas or tests exposed weaknesses, it had not always been possible to make the changes. The project was too far along. It could not be stopped to go back—not and still make a 1969 moon landing. So Apollo officials decided to build two Command Modules—the so-called Block 1 and Block 2 versions. Block 2, the polished version, would take men to the moon, while Block 1 would be used for early unmanned and manned check-out flights.

The Block 1 vehicles did not contain the docking tunnel through which astronauts would crawl to the Lunar Module because the Block 1s had been designed before the lunar-orbit rendezvous method of going to the moon was decided upon. They also lacked the more advanced hatches and several of the interior changes planned for Block 2. As Joe Shea, the spacecraft manager at Houston, described it, Block 1 was "a legacy of problems associated with getting the modes defined in the first place."

From November, 1963, through the end of 1965 there were ten unmanned launchings in which various Apollo systems, primarily the escape and abort mechanisms, were tested. In the first full-scale unmanned test, on February 26, 1966, the spacecraft, including the Command Module and the Service Module, was fired 5,500 miles down the Atlantic Missile Range by a Saturn 1-B, forerunner of the Saturn 5. The rocket engine so essential for lunar-orbiting maneuvers was fired and refired but never quite reached its peak thrust. Despite this, Apollo officials rated the flight a "successful first step."

Two more tests were run in the summer of 1966. On July 5 a dummy payload was put into earth orbit to prove the reliability of the rocket stage that would boost men out of near-earth orbit toward the moon. This was the engine S-4B that would be the third stage of the Saturn 5. The test was successful. Equally successful was a second launching on August 25, when an unmanned Apollo rose from Cape Kennedy on an 18,000-mile flight that ended in a Pacific Ocean splashdown. The spacecraft's engine fired perfectly and the heat shield withstood the temperature and pressure of a simulated reentry from orbit.

Soon thereafter, with the last of the Gemini flights concluded, all eyes focused as never before on Apollo.

Time of Adolescence

The spacecraft for the first manned mission was accepted by NASA in the summer of 1966 and shipped to Cape Kennedy. But engineers there uncovered a mare's nest of troubles. In October the oxygen regulator in the life-support system failed and had to be redesigned. Then the entire life-support system had to be returned to the factory for major changes. A leak developed in the cooling system, necessitating more time-consuming repairs. All hope of launching Apollo 1 in 1966 disappeared.

Shea professed not to be overly worried by the host of mechanical problems, referring to them as an expected part of the spacecraft's "maturation process." At a two-day briefing for newsmen in December, he disclosed that at least 20,000 failures had been logged in Apollo during the six years of its development. There had been more than 200 failures in the life-support system alone. "It is kind of like watching children grow up," Shea told newsmen in explaining Apollo's growing pains. "My feeling is we are somewhere in the middle of adolescence."

If the astronauts had any reservations about the space-craft, they did not voice them in public. In the afterglow of the Gemini successes, anything less than a safe and successful mission for Apollo 1 seemed remote.

Roger B. Chaffee, a 31-year-old lieutenant commander, was the youngest of the three Apollo 1 crewmen. Born in Michigan, he joined the navy after his graduation from Purdue in 1957 as an aeronautical engineer. Before becoming an astronaut in 1963, he had been a jet pilot. He may have been the rookie of the crew, but he did not lack self-confidence. "Hell, I'd feel secure taking it up all by myself," he once said. "You feel secure because you know what you're doing."

Chaffee felt good about the Apollo. "I think we've got an excellent spacecraft," he said in December. "I've lived and slept in it. We know it. We know that spacecraft as well as we know our own homes, you might say. Sure, we've had some developmental problems. You expect them in the first one."

Edward H. White was a 36-year-old lieutenant colonel in the air force who seemed destined for a place on one of the earliest lunar-landing flights, perhaps the first one.

Flying was in his blood. His father, a retired air-force major general, was an aeronautical pioneer who started out in army balloons, then switched to powered aircraft. His astronaut son was graduated from West Point in 1952, earned a degree in aeronautical engineering at the University of Michigan and in 1962 joined the space team. He was picked for the initial Apollo crew on the strength of his performance as the nation's first "space walker" during the Gemini 4 mission.

White always drove himself hard to achieve his goals, whether it was to star in soccer, set a West Point record in hurdles or run several miles a day—all of which he did. Apollo was his latest goal and he seemed to have no apprehension about being on the shakedown flight. "I think," he said, "you have to understand the feeling that a pilot has, that a test pilot has, that I look forward a great deal to the first flight. There's a great deal of pride involved, in making a first flight."

Virgil I. Grissom—everyone called him Gus—was a 40-year-old veteran of shakedown flights, having ridden the second Mercury and the first Gemini. Now the short, quiet air-force lieutenant colonel was the commander of Apollo 1. Grissom was born in Indiana, got a mechanical-engineering degree from Purdue and flew 100 combat missions in the Korean War. He went on to become a crack test pilot before being selected as one of the original seven astronauts. Though he lacked Glenn's charisma or Schirra's dash, he was a determined man, a careful engineer and a no-nonsense pilot.

Of the chances of a catastrophe, Grissom said in December of 1966: "You sort of have to put that out of your mind. There's always a possibility that you can have a catastrophic failure. Of course, this can happen on any flight. It can happen on the last one as well as the first one. So, you just plan as best you can to take care of all these eventualities, and you get a well-trained crew and you go fly."

That was that—until January 27, 1967.

Fire

On that Friday afternoon the three fully suited astronauts lay strapped in the couches of Apollo 1—Grissom

on the left, White in the middle and Chaffee on the right.
The spacecraft was perched on top of the Saturn 1-B, the
two of them standing a total of 224 feet tall. Around it
was the protective steel latticework of the 310-foot-high
service tower, which provided access platforms for workmen. Test conductors kept instrument watch on Apollo 1
from the "white room," the enclosed uppermost platform
nearest the spacecraft, and from a blockhouse about 1,000
feet away.

Apart from the rocket's not being loaded with fuel, it
was a realistic rehearsal of how the final countdown was
to go. It was not considered a "hazardous" test because
of the unloaded rocket. So everyone was somewhat relaxed
and there were no emergency crews of firemen and doctors
on hand.

With the astronauts inside the spacecraft, the hatch had
been closed and sealed and the cabin pumped full of pure
oxygen to a pressure of 16.7 pounds a square inch, 2 pounds
higher than normal sea-level atmosphere. This was the
way Apollo was to be pressurized while awaiting lift-off.

For the Apollo 1 crew it had been a frustrating day.
They had climbed into the spacecraft at about 1 P.M., a
couple of hours late because of some check-out problems.
Soon after they settled in, Grissom sniffed a strange odor in
his suit oxygen system. No impurities were found and the
practice countdown was resumed. Then there was a series
of communications problems—static and bad connections.
"How do you expect to get us to the moon if you people
can't even hook us up with a ground station?" an irritated
Grissom barked over the intercom. "Get with it out there."

At 5:40 P.M. a communications bug caused a new
interruption in the countdown, which had reached the
point where the spacecraft was supposed to switch from
external to internal power—that is, from electricity origi-
nating in the service tower to on-board electricity, simulated
in this particular test with storage batteries. If someone
could straighten out communications, the test would be
over in about ten minutes and the astronauts could, after
a debriefing, head for a relaxing dinner at some Cocoa
Beach restaurant.

The countdown was never picked up again.

At 6:30 P.M. the cabin pressure, cabin temperature and
oxygen-supply temperature were normal. There was a

brief surge of current, which was unexplained but of no apparent consequence. Sensors indicated that White's heartbeat and breathing rate shot up sharply, then returned to normal.

Suddenly it happened. At three seconds after 6:31 P.M. someone, either Chaffee or Grissom, yelled through the communications static something like "Hey" or "Fire." Some listeners believe they heard Chaffee report, "I smell fire."

A second later, monitoring devices on the ground indicated movement inside the spacecraft, then a rise in cabin temperature. The cry from inside became clearer and more insistent. "We've got a fire in the cockpit." It sounded like Chaffee.

More movement. Higher temperature. More garbled alarms. "They're fighting a bad fire . . . let's get out . . . open 'er up." Or "We've got a bad fire . . . let's get . . . we're burning up." The voice again seemed to be Chaffee's. A sharp cry of pain could be heard. Then silence.

At 6:31:17, or 14 seconds after the fire had first been detected, the cabin pressure reached a level of about 29 pounds a square inch and the cabin ruptured.

It took approximately five minutes for pad workmen to fight their way through the flames and acrid smoke to open the hatch. It was too late. The spacecraft's interior was a smoke-blackened, charred wreck. The pad crew found the limp bodies of the astronauts. White was lying in his couch with his arms over his head as though grasping for the hatch. Grissom also seemed to have an arm outstretched. Chaffee was slumped in his couch.

When doctors arrived, at about 6:45 P.M., the three were pronounced dead. Investigators later estimated that the men had lost consciousness about 15 to 30 seconds after the fire was first detected. The official death certificates listed the cause of death as asphyxiation due to smoke inhalation.

The Last Tapes

What happened? The question was to haunt all who worked on Apollo for the next weeks and months. At first even the sketchiest details, as reconstructed above from public documents released later, were not known.

General Phillips, who was in Washington, immediately ordered all data concerning the test impounded and took a special NASA plane for Cape Kennedy that night. Like many other reporters, I flew down to the Cape during the night to be on hand for Phillips's news conference the next morning.

Standing taut and ashen-faced, his back to Pad 34, Phillips announced that preliminary findings indicated that the three men had been killed by a "flash fire that originated in the spacecraft." The cause of the disaster, the first to befall astronauts in the line of space duties, was a mystery. They apparently died instantly, Phillips said, and never had a chance to escape. The only words they spoke, he said, were a quick message of trouble: "Fire in the spacecraft."

It was a difficult moment for Apollo, and NASA was totally unprepared for it. An information blackout was imposed. What information was released was often contradictory. NASA press officers in Houston and at the Cape could not, for example, agree on the simple fact of whether the spacecraft was on internal or external power at the time of the accident. Even Phillips was not sure. (External, it turned out.) Though a small point, it was indicative of the snarled lines of communication throughout NASA. Moreover, though Phillips said the astronauts had died almost instantly after one distress call, there were people at the Manned Spacecraft Center and at the plant of North American Aviation, the prime contractor for the spacecraft, who had tape recordings of their last seconds. They had played them over and over again on Friday night, the night of the tragedy, in search of possible clues. They were well aware that the men had spent some time fighting to get out. They knew there had been more than one cry for help, for one man in Houston used the word "gruesome" in describing the tape recordings in a phone call to NASA headquarters.

But NASA's highest officials insisted that they knew nothing of this important piece of evidence until after a story I wrote appeared in *The New York Times* on Tuesday, January 31. I bring this up because it was symptomatic of a more serious lack of candor that Congress and the public later found NASA to be guilty of.

Nothing so arouses a reporter's doubting instincts as a "no comment" or a vague and evasive reply to a question. And this is what reporters encountered on every hand throughout the weekend of the fire. They bitterly recalled Gemini 8. When the capsule went out of control, NASA withheld the tapes of communications during the crisis with the explanation that the astronauts' voice levels could give a false impression of their behavior. Naturally this led to speculation that the astronauts had panicked. Later, when the tapes (presumably the full tapes) were released, they showed that, on the contrary, the astronauts had handled themselves with workmanlike calm. Because of that episode, reporters took to referring to the NASA initials as meaning Never A Straight Answer. William Hines, a pugnacious NASA critic who was then on the *Washington Star*, observed: "I have found that while NASA has more information than a dog has fleas while all is going well, the experts tend to pull back for regrouping when things go wrong."

Faced with NASA's stony silence, I asked Douglas M. Dederer to poke around and see what he could find out. Dederer, part-time correspondent for the *Times* at Cape Kennedy, was a long-time resident of the area with many friends and contacts. His poking paid off late Monday afternoon. We found a highly placed engineer who had spent all day listening to the tape recordings of the astronaut's last seconds. After I heard what he had heard and I made a confirming phone call, I rushed to file a story that began as follows:

CAPE KENNEDY, Fla., Jan. 30—"We're on fire . . . get us out of here!"

These were the last words heard from the Apollo 1 astronauts who perished Friday when their capsule erupted in flames, an official source said today.

The shrill voice was believed to be that of Lieut. Comdr. Roger B. Chaffee of the Navy.

The astronauts' last moments, clocked at 16 seconds, were described by an engineer who spent most of the day listening to tape recordings of the fatal test and who heard reports from men on the launching pad at the time of the tragedy.

Commander Chaffee and two other crewmen, Lieut.

Cols. Virgil I. Grissom and Edward H. White 2nd of the Air Force, were scrambling, clawing and pounding to open the sealed hatch and escape the inferno in their Apollo cockpit, the source said.

But the hatch was unyielding, he said, and they had no time to reach for the ratchet that is normally used to unscrew it. There was no automatic release button.

The first hint of trouble, according to the source, came in almost casual tones.

"Fire . . . I smell fire," an unidentified astronaut reported over the intercom.

Two seconds passed.

"Fire in the cockpit!" cried Colonel White.

This time the voice was sharp and insistent. It was identified as Colonel White's by Donald K. Slayton, a former astronaut and now chief of crew operations.

There was silence for three seconds—then an hysterical shout from an unidentified astronaut:

"There's a bad fire in the spacecraft!"

A longer silence followed, about seven seconds. There were sounds of frantic movement, unintelligible shouting. Finally, after another four seconds, Commander Chaffee cried out the last words of distress: "We're on fire . . . get us out of here!"

It was previously reported by the space agency that the last words, from an unidentified voice, were "Fire in the spacecraft!"*

After the story broke, on page 1 of the *Times,* top NASA officials huddled in Washington to learn from their subordinates that there were indeed tapes and that the *Times* account was substantially correct. Three days later, Seamans returned from Cape Kennedy and released a short report confirming officially that the astronauts had made several distress calls, were moving around in the

---

* While the story certainly did not make me a candidate for any popularity prize in space circles, I have only two regrets. One is that it broke so that the grisly details were in print the day of the astronauts' burial ceremonies. The other is over some of the details that appeared in part of the story not reprinted here. Contrary to what I had been told, the astronauts' bodies were not burned beyond recognition. But insofar as the story may have led to a fuller disclosure of the facts of the fire, and of the Apollo project, I believe it was not only justified but useful. That the astronauts lived for several seconds after the fire began —as opposed to "instant death" in a "flash fire," as reported by NASA officials—indicated that they might have been saved if they had had a quick-opening hatch, which was incorporated in later Apollos.

cockpit and *did* let out "one sharp cry of pain." Later reports included an official transcript of the tapes that varied only slightly (due to the garbled communications) from what I had been told.

Tracing the Spark

One of NASA's first acts after the fire was to appoint a board of inquiry headed by Floyd L. Thompson, the white-haired director of the Langley Research Center and one of the most respected men in the aerospace profession. Known officially as the Apollo 204 Review Board,* the panel convened immediately, interviewed eyewitnesses and called in roughly 1,500 technical experts in fire and explosives, spacecraft design, chemicals, electronics and medicine. The investigators dismantled the cockpit in search of clues and examined all the recorded data of the fatal test. After ten weeks of intensive work, the board submitted a 3,000-page report that went beyond the fire itself to dissect the entire Apollo engineering and manufacturing process. The board strongly criticized those involved in the Apollo project for "many deficiencies in design and engineering, manufacture and quality control."

The board surmised that the fire broke out near Grissom's couch, probably from a faulty electrical wire. Under the couch was a bundle of wires next to a small door that led to a waste-disposal unit. "When the door was opened and closed," the board said, "its lower edges may have chafed the wires of this bundle, since there was no chafing guard for the Teflon insulation on these wires." The chafing may have torn the insulation, causing a wire to spark.

---

* The number refers to the manufacturing designation of the Saturn 1-B rocket for the spacecraft that burned. The other board members were Frank Borman, the astronaut; Maxime Faget, director of engineering development at the Manned Spacecraft Center; E. Barton Geer, associate chief of the flight vehicles and systems division at Langley; Colonel Charles Strang, chief of the missiles and space safety divisions of the air-force inspector general's office; Robert W. Van Dolah, director of the Explosives Research Center of the Bureau of Mines; George C. White, Jr., director of reliability and quality of the Apollo Program Office, and John Williams, director of spacecraft operations at Cape Kennedy.

Once ignited, the fire apparently swept through the cockpit from left to right, fed to furnacelike intensity by the pure-oxygen atmosphere and spread by combustible materials. Some nylon netting was one culprit. It was strung under the couches to prevent articles from falling into the equipment areas during testing. The netting, the board found, was the first substance to catch fire. Firebrands of molten nylon sputtered to other parts of the cabin. Other plastic materials caught fire. Under normal circumstances, they were highly fire-resistant, but in pure oxygen at high pressure they exploded into flames. Soldered connections in the aluminum oxygen pipes then began to melt, leaking more oxygen to feed the fire.

Death came to the astronauts, the board concluded, when the heat raised the cabin pressure to more than twice what it had been. Then the cabin ruptured. The fire, no longer supported by pure oxygen, began to smolder. But a burst of heavy black smoke and soot got into the astronauts' breathing apparatus and killed them through carbon monoxide asphyxiation. The burns themselves were not necessarily fatal.

The board said that it was impossible to determine the exact time the astronauts lost consciousness and when they died. But it was estimated that the cabin atmosphere became lethal 24 seconds after the fire started. "Consciousness was lost between 15 and 30 seconds after the first suit failed," the board concluded. "Chances of resuscitation decreased rapidly thereafter and were irrevocably lost within four minutes."

The tragic point was inescapable. The three men never had a chance. Why? "The Apollo team failed to give adequate attention to certain mundane but equally vital questions of the crew safety," the board charged. The astronauts were doomed by thoughtlessly placed wiring, haphazardly placed combustible materials scattered throughout the cabin, the pure oxygen they were breathing and a cumbersome escape hatch. Under the best conditions, it would have taken 90 seconds for them to release the three-layer hatch. Ironically, a quick-release hatch had been rejected for Apollo because of Grissom's Mercury experience when the spacecraft's explosive-release hatch opened prematurely after splashdown and almost drowned him.

In its report the board drew up a list of the contributing causes of the fire and it read like an indictment. They were as follows: a sealed cabin, pressurized with a pure-oxygen atmosphere without thought of the fire hazard; an overly extensive distribution of combustible materials in the cabin; vulnerable wiring carrying spacecraft power; leaky plumbing carrying a combustible and corrosive coolant; inadequate escape provisions for the crew, and inadequate provisions for rescue or medical assistance.

To prevent a recurrence the investigators recommended that NASA make an in-depth review of all spacecraft systems, restrict the use of combustible materials, improve rescue procedures, provide a quick-opening hatch and use an airlike atmosphere instead of pure oxygen during ground tests and possibly for orbital flight as well.

Placing the Blame

The report, like the fire, laid bare the weaknesses of the Apollo program as never before. From the president on down to the lowest-paid NASA engineer, the image projected had been one of near perfection—systematized, computerized and almost infallible. This facade was now shattered. One sentence leaped from the pages of the report into the headlines and the halls of Congress. The sentence: "The Board's investigation revealed many deficiencies in design and engineering, manufacture, and quality control."

Why had this been allowed to exist?

On April 10, the morning after the report was made public, the House Subcommittee on NASA Oversight— a title that had never before seemed so appropriate— wanted to know the answer. It was a bitter moment for James Webb, the tough, effusive optimist who had bulldozed NASA along the road to the moon so successfully until now. When he testified before the subcommittee, Webb was contrite and said that NASA accepted its share of the blame for overlooking design flaws in the spacecraft. But he reminded the congressmen, "Whatever our faults, we are still an able-bodied team."

J. Leland Atwood, president and chairman of the board of North American Aviation, testified that his engineers also accepted a share of the blame for underestimating the threat of fire. Atwood defended the company's standards of production, inspection and quality control. But under questioning by congressmen he indicated that "overconfidence" as a result of previous space successes might have led to some relaxation of standards. "We recognize that while we have made every effort to avoid any deficiency," Atwood conceded, "some deficiencies did in fact exist."

As the hearings progressed, friction developed between NASA and North American. Their solid front of shared blame broke. North American's Dale D. Myers, the Apollo program manager, complained that NASA-inspired modifications as late as 1966 had added 70 percent more wiring to the spacecraft and made it impossible to complete and check it before it left the factory. NASA's Seamans testified before the Senate Aeronautical and Space Sciences Committee, which had opened a parallel investigation, that "North American had not always shown sufficient dedication to the engineering design or workmanship on the job." As Evert Clark of *The New York Times* observed, "Each partner is now reaching into the file cabinets for carefully preserved records that would tend to absolve itself and shift the blame to others."

The Phillips Report

This inevitably lifted the lid off several matters that NASA and North American would have preferred to keep from public debate. Under congressional questioning it was disclosed that more than a year before the fire NASA had grown so dissatisfied with its prime contractor's performance that thought was given to taking part of the job away. Samuel Phillips went to North American's plant in Downey, California, to determine why the contractor was unable to meet its delivery schedule and stay under cost ceilings. What resulted came to be known as the "Phillips Report."

Phillips insisted that it was not a formal report, only a "set of notes." Whatever the form, in content it was a hard-nosed appraisal of North American's performance. After his trip to Downey, Phillips sent a letter to Atwood on December 19, 1965, in which he said NASA was unhappy with the progress on the spacecraft and on the second stage of the Saturn 5 rocket, another North American project. The letter was not made public. Nor was Congress informed, though some allusions were made in testimony. In the spring of 1966 NASA advised a House space subcommittee that Apollo schedules were being met. There was no hint of the Phillips findings.

The fact that they had not been told seemed to bother congressmen as much as what they heard about Apollo problems. When they pressed for details, Phillips at first denied that there was a "Phillips Report." In his own mind, it was not a report. Webb was reluctant to discuss the matter because, he said, it would destroy the "intimate and confidential" nature of dealings between NASA and its contractors. Finally, however, NASA released a summary of the Phillips investigation to the House subcommittee.

"Our main criticism," according to the summary of what Phillips told Atwood, "was that [North American] was overmanned and that the S-2 [second stage] and CSM [Command and Service Module] programs could be done, and done better, by fewer total people, better organized with particular emphasis on achieving greater competence in key management and technical positions."

Phillips said that he had found that corporate-level interest in the Apollo programs at North American Aviation was "passive." "There is no evidence," he reported, "of current improvement in NAA's management of these programs of the magnitude required to give confidence that NAA performance will improve at the rate required to meet established Apollo program objectives."

The report also pointed to "inaccurate cost estimates," "lack of planning," "late or incomplete engineering" and a "large number of discrepancies discovered by both North American and NASA inspectors."

Phillips gave North American until April, 1966, to get its house in order—or else. He was not clear in his ultimatum, but there was an implied threat of a complete rewriting of the contract. But by April, NASA said, North American had begun to correct many of its shortcomings and improve workmanship. Therefore, no action was taken.

Dark Shadows

To Congress—and to the public—this raised certain questions. If Phillips was wondering whether North American was qualified to carry out the Apollo contract, should the company have gotten the contract in the first place? And, by the way, how *did* North American get the contract?

So the lid opened even further, exposing some dark shadows in Apollo's background. The Senate committee, in particular, pressed hard for answers.

North American had been selected in November, 1961, from among five aerospace companies. The four others were General Dynamics, General Electric, the Martin Company and McDonnell. Their proposals were submitted to a panel of space agency experts known as the source evaluation board. This board ranked Martin highest overall in its technical approach, technical qualifications and business qualifications. But North American was chosen. The reasons had never been fully explained.

When Congress wanted to know, Seamans drew up a memorandum, dated June 9, 1967, which was submitted to the Senate space committee. According to the memorandum, the source evaluation board had reported its choice of Martin to Webb, Seamans and Hugh Dryden, who died in 1965. They were the top three administrators at the time. In spite of the recommendation, the NASA triumvirate selected North American, contending that it "had by far the greatest technical competence." What appeared to be a clincher was North American's greater experience in building vehicles for manned flight, especially its work on the X-15 rocket plane.

According to the memorandum, Webb, Dryden and Seamans had "unanimously agreed" that:

    A. Both Martin and North American Aviation have good business capability. Either company would be qualified to carry out the business management responsibilities on Apollo.

    B. Martin had the best technical approach; but the advantage the company enjoyed through having had an Apollo study contract diluted this apparent superiority.

    C. North American Aviation had by far the greatest

technical competence; and that because of the very
great engineering complexities involved in the Apollo
project, this competence had to be considered as an
overriding consideration in selecting a contractor.

D. North American had submitted the lowest cost
proposal and has an excellent reputation for low-cost
operations, as well as the best cost history of any of
the companies involved.

E. North American Aviation, Inc., Space and Information Division, should be selected as the company
to develop and produce the Apollo spacecraft.

These disclosures took Congress by surprise. When
Webb reappeared before the House subcommittee, he encountered harsh criticism for NASA's apparent lack of
candor in the past. "We have had to take you at face
value," Representative John W. Wydler of New York
commented. "Maybe it was our fault for taking you at
face value." Reflecting the attitude of most legislators, Representative Ken Heckler of West Virginia warned: "I intend to be much more skeptical of NASA in the future, on
this program and others."

Skepticism was understandable. The explanation of how
North American had obtained the Apollo contract left
something to be desired. What of this "excellent reputation
for low-cost operations"? North American had said the initial phase of its contract would cost in excess of $400
million. But by the time of the fire the amount had risen to
$2.2 billion. Part of this, of course, was due to NASA
alterations, but not all. And what of the judgment that, according to the memorandum, North American's key personnel were "better than Martin's and by a significant degree"? After the fire, during the storm of controversy,
North American found it necessary to hire two of Martin's
top men from previous space programs—Bastian "Buzz"
Hello to head its Cape Kennedy operations and William B.
Bergen to head the entire space division. Moreover, why
penalize a company like Martin for having had the advantage of prior study of the project and submission of preliminary proposals?

The circumstances of the contract award led several reporters to take a closer look at something else that was going on in Washington in 1961, the Bobby Baker Case. On

April 17, 1967, Harry Schwartz, a member of *The New York Times* editorial board, considered Baker's possible role in Apollo in a piece for the editorial page. It read in part:

> It is worth exploring the granting of the contract for another reason. For there is evidence to suggest that the shadow of Bobby Baker, the former secretary to the Senate's Democrats, darkened the history of North American's relations with NASA.
>
> During the period when NASA was pondering the choice of contractors for the Apollo Project, North American was holding talks that ended in the Serv-U Corporation acquiring the highly profitable contract for vending machines in North American's plants.
>
> Serv-U, of course, was the company owned by Mr. Baker and Fred R. Black Jr., North American's highly paid Washington lobbyist. They were able to start Serv-U as the result of a loan from a bank that had a close financial relationship with Mr. Baker's good friend, the late Senator Robert S. Kerr, then chairman of the Senate Space Committee and a great booster of Oklahoma's industrial development. And at the head of NASA was—and is—James E. Webb, who also was a close business associate of Senator Kerr before he took his present post.
>
> Against this background it may not be surprising that after North American got the huge Apollo contract from NASA it picked Tulsa, Okla., as the site for a new plant and did much of its banking with a bank in which Senator Kerr and Mr. Black had substantial amounts of stock.
>
> Every defense contractor has friends in high places and there has been no suggestion that these particular relationships influenced the choice of North American. But a full investigation of the Apollo project from contract award to the present seems called for in order to insure that so avoidable an accident [the Apollo fire] does not happen again.

There was, however, no evidence to suggest Baker and Black had exerted any illegal influence. In a conversation several months later, I asked Webb about the persistent rumors of possible under-the-table deals involving Apollo.

"There were absolutely no political implications in any of our contract decisions," Webb stated.

Picking Up the Pieces

As the congressional hearings ground to an end, everyone involved with Apollo was all but paralyzed by the crisis of confidence gripping the project. Active training of astronaut crews was suspended. Morale slipped lower and lower. In a slap on the wrist for North American, the Boeing Company was brought in to assist in the management of the project and signed a contract to perform many "technical and evaluation tasks" for future Apollo flights. And Joe Shea, the spacecraft manager for NASA, was shifted from his job to a new position in headquarters.

Shea personified the great depression into which the project had fallen. The death of the astronauts had shaken him personally as well as professionally. He had known these men and played handball with them. He felt directly responsible for the fire—a burden that no one man was asked to bear or should have tried to bear. Into Shea's place moved George M. Low, who was deputy director of the Manned Spacecraft Center and who had been a NASA engineer and administrator from the beginning.

The fire shook many others besides Shea. When Bergen became president of North American's space division, replacing Harrison A. Storms, he said: "I found some of the people in almost a state of shock. North American had had such a good reputation over the years that they couldn't understand what could possibly have so suddenly happened."

Slowly NASA and North American patched up their differences and set about to get Apollo moving again. Engineers speeded up work to build a quick-opening hatch, something that had already been designed for Block 2 Apollos. New fire-resistant spacesuits, made of a glass fiber called Beta cloth, were ordered. Nylon netting and other inflammable materials were replaced with more fire-resistant plastics. Mock-ups of Apollo underwent severe fire tests in laboratories. The pure-oxygen environment was retained, but changes were made to provide an added measure of safety during ground testing. Workmen, so-bered by the fire and the bad publicity, exercised a new meticulousness, inspectors a new care. "Some people are literally killing themselves, often working 18 hours a day," Rocco A. Petrone, director of launch operations at Cape Kennedy, said in discussing the back-to-business attitude.

Plans were laid for a new flight, with Schirra, Walter Cunningham and Donn Eisele as the crew. As the people of Apollo picked up the pieces, the memory of Grissom, White and Chaffee was real and haunting. But what the Apollo people remembered more and more was not the astronauts' deaths but something Grissom had said. "If we die," he once remarked, "we want people to accept it. We are in a risky business, and we hope that if anything happens to us it will not delay the program. The conquest of space is worth the risk of life."

There would be delay, but there was still a chance of meeting that 1969 deadline for a lunar landing. It would, however, be nearly two years—two long, trying years—before Apollo recovered from the shock and regained confidence in itself.

Chapter VIII

The Russians Also Fall Behind

With Apollo stalled, the advantage in the race to land men on the moon seemed to pass back to the Soviet Union. But could the Russians seize the opportunity?

This question was very much on the minds of those of us who had followed the contest since the Gagarin flight and the Kennedy speech. There were guesses and theories and many Americans feared that their chances of beating the Russians had gone up in the Apollo smoke—but there were no answers. For Soviet preparations and intentions had seldom been more veiled in mystery than in the spring of 1967, and no one could foresee that tragedy was about to strike an equally swift and devastating blow to Soviet space efforts.

Unaccountably two years had passed since cosmonauts had last ventured into space—in Voskhod 2 in March, 1965. In that time the United States had completed all the rendezvous flights of Gemini to establish leadership in manned missions. Had the nation that was first in space withdrawn from the race? Were the Russians having technical problems? No one could be sure—and the Russians weren't telling.

In fact, if asked, leaders of both the American and Soviet space programs would deny that there even was a race—it was just two aggressive efforts aimed at exploring space. But no American could seriously deny the catalytic influence of Gagarin's flight on the whole concept and course of the Apollo project. NASA officials found Congress more willing to loosen the purse strings in a year when the Russians seemed particularly threatening. Likewise, one suspects that the Russians were not infrequently influenced in their space spending by the news out of Cape Kennedy.

But Anatoly A. Blagonravov, the Soviet artillery general and rocket pioneer who often spoke for the Soviet space establishment, characterized the competition in the following way:

> Of course, both nations are deeply gratified by every new achievement. Of course, the Soviet Union takes just pride in the fact that the first artificial satellite of the earth was made and orbited by the Soviet people, that the first man to go into orbital flight was a Soviet citizen. These successes enhance the prestige of Soviet science and technology and also the prestige of our country generally in the eyes of the whole world. In this respect, the rivalry between the USSR and the USA does play a definite role, but it is not, by far, the mainspring behind the great expenditures for space activities. The main driving force of this research is the needs of science and concern for the future benefits to mankind.

To the Russians, though, the pride and prestige derived from space exploits were no small consideration, especially when such feats called attention to the great Soviet accomplishment in building a powerful nation out of peasant backwardness and the ruin of revolution and war. For this reason the ebullient Nikita S. Khrushchev, who as premier early in the decade was an enthusiastic booster of spaceflight, could not resist gloating over the successes. "Bourgeois statesmen," Khrushchev told a Polish audience in 1963, "used to poke fun at us, saying that we Russians were running around in bark sandals and lapping up cabbage soup with those sandals. They used to make fun of our culture, the culture of a people considered, so to say, to be the last among the civilized Western countries. Then suddenly, you understand, those who they thought lapped up cabbage soup with bark sandals got into outer space earlier than the so-called civilized ones."

It was this kind of up-from-inferiority-complex thinking that was expected to keep the Soviet Union driving toward the moon. It was generally assumed that the Russians were using the two-year hiatus to design, build and test a brand-new spacecraft. Perhaps this would be the Russian moon vehicle.

Vostok and Voskhod

Until 1967, cosmonauts had gone aloft in two classes of spacecraft, the Vostoks and Voskhods. There were six manned missions in a Vostok (two each in 1961, 1962 and 1963) and two flights in a Voskhod (one each in 1964 and 1965). But such was the secrecy of the Soviet program that not until now did Westerners get a fairly clear picture of these vehicles and their launching rockets.

The one-man Vostok was large and spherical and weighed roughly 10,400 pounds, about 3½ times as much as the Mercury. Much of this extra weight, however, went not into considerably more furnishings and systems than Mercury had but into Vostok's outer wall of solid steel alloy. Vostok was roomier than Mercury. Its diameter was about eight feet. There were two compartments, one for the pilot and one for the equipment. Those who had been inside a Vostok cockpit described it as much simpler than Mercury's, apparently containing fewer dials, gauges and switches. One major difference was the cabin atmosphere. Instead of the low-pressure pure oxygen that American astronauts breathed, the Vostok cabin was pumped full of mixed gases (24 percent oxygen and the rest primarily nitrogen) to a pressure almost like that at sea level. In short, it was almost identical to the air we ordinarily breathe. The stouter walls of Vostok made it possible to maintain such a pressure, which might have ruptured the thin Mercury walls. Like the American craft, Vostok was equipped with braking rockets and parachutes for the return to earth. But it was designed so that when it got near the ground the hatch opened automatically and, moments later, the pilot in his seat was explosively ejected. He descended to earth by parachute as the capsule dropped under a separate chute.

To launch Vostok the Russians employed a liquid-fuel rocket assumed to have been a modified intercontinental ballistic missile. With a boosting power of 1,340,000 pounds, it had four times the thrust of the American Atlas used in Mercury. For many years, until the advent of the Saturn and Titan 3-Cs, the Vostok rocket gave the Soviet Union a crucial edge over the United States in weight-lifting capability. Its main stage was a cluster of four large engines strapped around a core engine. Each of the five basic units, though equipped with a single pair of fuel pumps, had four exhaust nozzles. Thus during full thrust the rocket would spew exhaust from 20 openings. Then the core continued to fire for a short time after the strap-ons shut down. The rocket's main upper stage had a single engine. But it was a small third stage that actually placed the Vostok into earth orbit.

For cosmonauts the point of departure into space was a launching complex east of the Aral Sea. Russians call it the Baikonur Cosmodrome, after the town of Baikonur. However, the exact site is much nearer the village of Tyuratam, some 230 miles southwest of Baikonur. This is the Soviet Cape Kennedy. In addition, there are two other launching complexes. At Kapustin Yar, on the Volga River, the Russians launch many of their smaller satellites, fire sounding rockets for upper-atmosphere studies and conduct some missile tests. Around 1966 the Russians began launching also from an area near the town of Plesetsk, south of Archangel. From this far-north site they send up satellites into orbits crossing the poles instead of following roughly equatorial paths.

The Russians shifted from the Vostok series to the multimanned Voskhod flights after they developed an improved main upper stage—one with twin engines—that made it possible for them to generate a thrust of 1,433,250 pounds with the otherwise unchanged Vostok rocket. At first Americans assumed that Voskhod was an entirely new spacecraft. Still smarting from the Sputnik surprises, they tended to credit the Russians with greater space capabilities than they actually possessed. Subsequent analysis, however, led Western space experts to conclude that the Voskhod was nothing more than a Vostok with modifications. The same eight-foot sphere was rearranged to accommodate three men sitting side by side. The first manned Voskhod, sent aloft in October, 1964, weighed 11,731 pounds and carried a crew of three. Russians contended that Voskhod had an improved braking rocket for gentler landings which obviated the need for parachute jumping by the crew just before touchdown.

The second Voskhod, launched five months later, was heavier, weighing 12,529 pounds. It was heavier, although it carried two cosmonauts instead of three, because it included a submarinelike airlock. It was through this chamber that Aleksei A. Leonov crawled to begin man's first "walk" in raw space. When it came time for reentry from orbit, Voskhod 2's automatic positioning system failed to work properly. The cosmonauts had to wait another orbit and then steer the ship down manually. They alighted safely, but way off the mark. Instead of landing in the open country of the Ukraine, where recovery teams were waiting, they came down far to the north in a snowbound forest. It took hours for the ship to be located and about a day for ground parties to break through the wolf-infested woods to reach the crew and bring them out on skis—a rather inglorious return for the world's first space walker.

Confusion of Purpose

Thus ended the rather brief Voskhod series—and all Soviet manned spaceflight for more than two years. During the hiatus the Russians achieved unmanned landings on the moon, the orbiting of a mammoth spacecraft called Proton and launchings of a host of vehicles, some for science, some for military surveillance and some for testing new equipment, but all lumped under the unenlightening label of Cosmos. This showed that they were still active in space; in the total weight of their payloads they maintained a superiority. But no one could be sure what direction all this activity was taking. The Russians' persistence in flying unmanned probes to the moon indicated a strong interest in lunar exploration. Yet they had not tested a rocket powerful enough to send a manned craft all the way to the moon. The Protons, weighing more than 13 tons each, were taken to be possible building blocks for large manned earth-orbiting laboratories of the kind the Soviet Union professed to be planning. But assembling the components of such laboratories required an ability to rendezvous and dock in space, something the Russians had yet to demonstrate.

So what did all the unmanned activity add up to? In October, 1966, I sought the answer to this question at the 17th annual International Astronautical Congress. A few dozen Soviet delegates were among the 1,000 scientists and engineers from 30 countries at the six-day gathering in Madrid.

The Soviet scientific papers were unrevealing, dealing mostly with biomedical aspects of flight and some rather obscure mathematics. In conversation the Russians replied evasively when asked about their plans for the future. In an interview, however, Leonid I. Sedov, head of the Soviet delegation, conceded to me that several "difficult problems" had to be overcome before cosmonauts could attempt a trip to the moon. He said that the problems involved operations to place men in a lunar orbit, to land them safely on the lunar surface and especially to bring them back to earth. His remarks reminded me of what Khrushchev had once said more succinctly. The question, Khrushchev had declared, was "not mooning man, but de-mooning him."

Sedov's comments, vague as they were, indicated that the Russians were none too confident of their ability to send men to the moon any time soon. Westerners I talked to at Madrid believed that the Soviet Union was in the midst of developing an entirely new, larger and more maneuverable craft, but was grappling with some serious technical snags. One problem may have been the loss of Sergei Pavlovich Korolev, chief designer of Soviet space vehicles and rockets.

The Chief

Little was known of the Soviet space establishment— where its research centers and factories were, how much money was budgeted, who its leaders were. Why? Khrushchev had once said, "Enemy agents might be sent in to destroy these outstanding people, our valuable cadres."

Only his death of cancer in January, 1966, at the age of 59, liberated Korolev from imposed obscurity. The man who never received any public acclaim in life was given a hero's burial in Red Square. For it was he, the chief designer, who had directed the launching of the first Sputnik, designed the Vostok and Voskhod vehicles and planned the various mooncraft. In addition, he was credited posthumously with having played a major role in designing the rockets for the manned flights. In that case,

Korolev must have exerted a greater influence over the Soviet space effort than any single man had in the American program. According to Gherman S. Titov, the second cosmonaut to orbit earth, Korolev was "the single most important man in the Soviet Union's cosmic flight program—a man who walks with the giants."

After Korolev's death, the job of chief designer apparently passed to a Ukrainian engineer named Mikhail K. Yangel, of whom little was known and nothing was reported by the Russians. With the job Yangel inherited the secret blueprints of Korolev's newest creation—a spacecraft to be called Soyuz, pronounced Suh-Yoosh.

In early 1967 tantalizing hints of such a new-generation vehicle and of impending flights began to circulate. In speeches and interviews cosmonauts and scientists talked ambitiously of manned voyages to the moon and of an advanced craft capable of carrying "more than five passengers." At the Madrid meeting Oleg G. Gazenko, the leading Soviet expert in space medicine, had said that his country was "doing very serious preparatory work" for another manned mission that would be a "serious new step" in space exploration.

Death in Soyuz

The long-awaited Soyuz was introduced to the world on April 23, 1967. After several unmanned test flights, it was launched from Tyuratam with one man aboard—Colonel Vladimir M. Komarov, a veteran of the Voskhod 1 mission and the first cosmonaut to go aloft a second time.

The first announcements of the flight by Tass, the Soviet press agency, were terse. By the fifth orbit Tass merely said that the flight was "proceeding according to plan." Since no description was given of the craft or its mission, Westerners could only speculate. The most popular guess was that Soyuz 1 would be joined in a day or so by a second Soyuz for rendezvous and docking maneuvers. The word Soyuz means Union, which lent support to the idea. The guess also jibed with the often-expressed interest of the Soviet Union in orbiting large space stations, which would require experience in rendezvous and docking.

No launching occurred on the second day. Perhaps something went wrong in the countdown for the second ship, if there was one. Tass said on the morning of the 24th that Komarov was in good health and "the ship's systems are functioning normally." Then, to nearly everyone's surprise, Komarov started his return to earth after only 17 orbits.

In a story two days later, Raymond H. Anderson, Moscow correspondent of the *Times*, said that a Soviet reporter who was at the Tyuratam space center had given the following account of the last minutes of the Soyuz 1 flight:

> "Well done!" the ground controller declared after he had heard that the retro-rocket system had functioned perfectly following the positioning of Soyuz 1 for re-entry.
>
> "Everything is working fine," replied the faint voice of Komarov, who was somewhere over Africa. "Everything is working perfectly." He seemed to be trying to reassure us.
>
> I did not hear his voice again. Several minutes passed, as long as a human life.

Komarov was dead, the Space Age's first casualty in flight.* According to Soviet reports, he died when the Soyuz plunged through the atmosphere and, parachute lines hopelessly tangled, crashed into the ground. Tass declared that the Soyuz had performed normally in orbit and had been successfully braked with retro-rockets. "However," Tass announced, "when the main parachute was opened at an altitude of seven kilometers [4.3 miles], the lines of the parachute, according to preliminary information, got snarled and the spaceship descended at great speed, which resulted in Komarov's death."

Many Russians broke into tears when the news was broadcast. The tragedy was a grievous shock to the Soviet people, who had looked to the flight to help them regain their space leadership. There had been talk of a

---

* Contrary to persistent rumors, Western observers can find no evidence of other flight casualties that the Soviet Union might be keeping secret. There were at least two known deaths of cosmonauts in training, one in a parachute mishap and another in an aircraft accident. In 1968 the most famous of the cosmonauts, Yuri Gagarin, and a backup cosmonaut, Vladimir S. Seryogin, were killed in a plane crash.

whole series of Soyuz missions throughout the summer. But now that Komarov was dead Soyuz, like Apollo, was going back to the drawing board. An investigation was ordered into "all aspects of the flight."

This implied that Soyuz 1 had encountered more problems than merely tangled parachute lines. There were rumors of serious trouble with the stabilization system, of the spacecraft's having been out of control in orbit. Charles S. Sheldon II, acting chief of the Science and Policy Research Division of the Library of Congress, observed in a report prepared for the House Science and Astronautics Committee that "if Komarov had planned to come down routinely, he would have done so most likely on the 16th or 17th orbit when he would be lined up with the normal recovery area." Instead, he came down at the beginning of the 18th orbit. This suggested, Sheldon said, that Komarov "came down earlier than originally planned because of troubles appearing soon enough that a second launch was cancelled" and "severe enough that he could not get down at the optimum pass." The Russians insisted, however, that Soyuz 1 had completed its mission and that no complications, other than the parachute trouble, had occurred.

In any event the Russians, stunned and disappointed, became even more reticent about their space program. At the time of the accident I had a request in for a visa to go to the Soviet Union and visit some of its scientific installations, perhaps even talk to its cosmonauts and space officials. In late May, a few days before my plane was to leave, the Soviet consulate in New York notified me that my visa request had not been approved. No explanation was given in response to my inquiries or those made in Moscow by the *Times*. It was generally assumed that the Russians, like the Americans after the Apollo fire, wished to say as little as possible about their failures.

The Look of Soyuz

More than a year later, toward the end of 1968, the Russians began to disclose some descriptive details of the Soyuz spacecraft. The drawings that appeared in Moscow newspapers showed a three-compartment vehicle with winglike solar panels extending from the midsection. The Soyuz had two separate crew cabins, instead of one, as in Apollo. In addition, there was a rear unit housing instruments and rocket engines, much like the Apollo's Service Module. During launching and return to earth and such key maneuvers in space as rendezvous with another ship, the cosmonauts would be seated in the heavily shielded spherical compartment in the middle. At other times they would live and work in the roomier compartment in the nose.

Altogether the two cabins were reported to have 318 cubic feet of space for the men to move about in—compared to 210 cubic feet in the Apollo crew cabin, which was about as roomy as the inside of a large station wagon. According to Soviet reports, the Soyuz could stay in orbit as long as 30 days with a single cosmonaut aboard or about ten days with a three-man crew.

Nothing the Russians said about Soyuz indicated clearly what role it was to play in exploration of the moon. Western observers suspected that a modified version, perhaps boosted by a more powerful rocket or assembled in earth orbit, could serve as the nucleus of a Soviet manned mooncraft. But the Russians, who had not spelled out their space goals and timetables, continued to be enigmatic about their lunar plans.

To Westerners they seemed to have three possible ways of going to the moon. By the first method they could, with a moderately powerful rocket of the Vostok class, send a mooncraft aloft in sections and assemble the sections as they orbited the earth. Once put together, the craft would be manned by means of a ferry vehicle and then fired out of earth orbit to the moon and back. By their continued talk of space-station assembly in orbit, the Russians encouraged this kind of speculation. The second method would involve lunar-surface rendezvous. The Russians would first launch an unmanned vehicle to the moon and soft-land it. It would have supplies and a return propulsion system on board. Once it was checked out by remote control, a second vehicle with the cosmonauts aboard would be sent directly to the moon. Since neither ship would have to be nearly as heavy as Apollo, no rocket the size of Saturn 5 would be required. Landing close to the first spacecraft, the cosmonauts would explore the lunar surface, then board the first spacecraft, fire its rocket and return to earth. The third method would probably be the simplest but it would require an enormous boost capability—on the order of 10 million or more pounds of thrust. With that kind of power the cosmonauts could voyage to the moon and back without any intervening rendezvous steps. Rarely was it speculated that the Russians planned to follow the Apollo approach of lunar-orbit rendezvous with a detachable landing craft.

Some Westerners suggested that the Russians might not even be aiming for the moon. Perhaps, they said, it was a one-sided race—and therefore no race at all. It was possible that the Russians, believing their chances of beating Apollo to the moon were slim, had shifted their emphasis to other targets—the massive earth-orbiting laboratories they talked of so much or the planets Venus and Mars.

But in 1967, the tenth anniversary year of Sputnik, no one was willing to count the Russians out of the moon race. For them, as well as for Americans, it was a year of tragedy and delay, but not withdrawal from the new frontier of space.

Chapter IX

Is This Trip Necessary?

No great and distant goal is reached without the determination to overcome setback and wavering resolve. Magellan, Drake and Sebastian Cabot were but a few of the seafaring explorers who had to quell mutinies to make their places in history. Others kept going in spite of shipwreck, scurvy and desertion. A much different test of resolve confronted Louis Antoine de Bougainville and his crew in 1768 when, in their search for the mythical southern continent, they stopped off at Tahiti. A young Tahitian woman came aboard the ship and, as Bougainville wrote in his log, "negligently allowed her loincloth to fall to the ground, and appeared to all eyes such as Venus showed herself to the Phrygian shepherd." It was 13 presumably delightful days before Bougainville could get his crew to resume the voyage. With no such respite, Columbus's men were a muttering lot a few days before their historic landfall. They found themselves in the seaweed-matted waters of the Sargasso Sea, the broad pool of near-stagnation that is the eye of the North Atlantic's rotating currents. They had come too far to turn back, but feared that they were stuck and would never get through to their destination.

In the summer and fall of 1967, Apollo was adrift in its own Sargasso Sea. Too much money and effort had already been expended on the enterprise to turn back. But the accident on Pad 34 had raised too many doubts about the project's management and engineering for anyone to feel assured that President Kennedy's goal of a landing in the decade would be fulfilled. With a little over two years to go, none of the major hardware—the Saturn 5, the Lunar Module or the Apollo Command Module—had yet been tested by men in flight. Another disaster could ruin whatever chances were left to make it to the moon in 1969. Indeed, another disaster, or another round of congressional and public criticism, might even kill the whole project.

It was not so much the 241,000 people then working on Apollo whose resolve was diminished. In fact, after the initial shock of the fire wore off, they seemed more determined than ever to press on. They sought to prove themselves. It was, instead, the American public that was having serious second thoughts about the 1961 Kennedy decision. (If they had had any way of knowing, Queen Isabella and her subjects might have had similar midocean doubts about Columbus.) Public opinion polls indicated that never before had Americans been filled with so many misgivings about the moon venture.

At a meeting of aerospace engineers in Seattle in August, 1967, I heard Representative Joseph E. Karth, a senior Democrat on the House space committee, lay it on the line. "Of all the thousands of letters arriving in Congress every week," Karth told the engineers, "none that I know of has asked for an increase in the space budget; few have opposed cuts. In fact, a substantial number are suggesting cuts in total spending and pinpoint space as the likeliest candidate."

A few weeks later, one of the country's outstanding aerospace industrialists put it this way. "The emotional reaction, the honeymoon with space is over," Simon Ramo, vice chairman of TRW, Inc., said. In a panel discussion at the annual meeting of the American Institute of Aeronautics and Astronautics in Anaheim, California, Ramo said that the average citizen now looked at the Apollo project and thought: "If we can provide good air for a man to breathe on the moon, why not on the streets of our principal cities? If an orbiting man can traverse the entire globe in a few hours, then why can we not apply technology in such a way as to get from home to work or to the airport in minutes rather than hours?"

Put still another way, people were more deeply disturbed by different things in 1967 than they were when the Apollo decision was made six years earlier. A growing number of Americans began to look upon Apollo as a magnificent irrelevance.

The Case Against Apollo

Apollo's critics usually raised three basic objections: (1) It cost too much. (2) The money and talent could be more usefully directed to down-to-earth problems. (3) It was a "childish stunt" to make a race out of going to the moon and insisting on such an artificial deadline as the end of the decade. In short, the moon was not worth the price or the rush.

About the only objection no longer heard in 1967 was the one about it being impossible to fly men to the moon in the foreseeable future. Even the fundamentalist complaints —man has no business going to the moon, or similar latter-day variations of the if-God-had-meant-for-man-to-fly-He-would-have-given-him-wings school of thought—were dying echoes. Arthur C. Clarke, the British science writer, reminded us in his book *The Promise of Space,* published in 1968, that every revolutionary idea—in science, politics, art or whatever—seems to evoke three stages of reaction. The stages, Clarke said, may be summed up by the phrases: "It's completely impossible—don't waste my time"; "It's possible, but it's not worth doing," and "I said it was a good idea all along." At the moment, Clarke added, "astronautics is still passing through stage 2."

Apollo's cost was an old and understandable subject of concern. The Republican party attempted to make a campaign issue of it in 1964, taking their cue from former President Eisenhower, who was never too sympathetic to space spending. Eisenhower had declared in 1963: "I have never believed that a spectacular dash to the moon, vastly deepening our debt, is worth the added tax burden it will eventually impose upon our citizens."

The issue was revived in 1967 and debated in many an American home as taxes rose, as the war in Vietnam imposed new strains on the federal budget and as the nation's long-neglected social ills became more evident. Yet, these conditions notwithstanding, never before had the American taxpayer been asked to contribute so much to scientific exploration as with the Apollo project. Bases in Antarctica, oceanographic ships and even atomic accelerators were trifling expenses by comparison. In the past, moreover, governments generally promoted exploration that promised to enlarge their territorial claims. Flag raising on newly discovered lands was traditional. But, by a treaty with the Soviet Union, no such claims were to be made on the moon. Nor was the moon expected to provide any ready source of wealth; the cost of transporting moon resources back to earth would be prohibitive for decades to come.

To drive home the enormous cost of the moon race, Warren Weaver, a mathematician and former president of the American Association for the Advancement of Science, once drew up a list of comparisons. With the sum of $30 billion, which he estimated Apollo would cost, Weaver said:

> One could give a 10 per cent raise in salary, over a 10-year period, to every teacher in the United States from kindergarten through universities (about $9.8 billion required); could give $10 million each to 200 of the better smaller colleges ($2 billion); could finance seven-year fellowships (freshman through PhD) at $4,000 per person per year for 50,000 new scientists and engineers ($1.4 billion); could contribute $200 million each toward the creation of 10 new medical schools ($2 billion); could build and largely endow complete universities with liberal arts, medical, engineering and agricultural faculties for all 53 of the nations which have been added [up to that time] to the United Nations since its original founding ($13.2 billion); could create three more permanent Rockefeller Foundations ($1.5 billion); and one would still have left $100 million for a program of informing the public about science.

Weaver's shopping list was often cited to support the argument that a trip to the moon was an extravagant diversion of national resources at a time when millions of people still lived like animals, disease was rampant, and school plants and teachers' salaries were inadequate. Kenneth B. Clark, a Negro psychologist at the City University of New York, told a congressional committee: "I just don't think the moon is going to be an adequate substitute for the fact that we haven't addressed ourselves to clearing up the slums." Senator J. William Fulbright suggested that the course of history was more likely to be influenced by how the United States dealt with the unemployed than the unexplored.

Looking at Apollo in strictly scientific terms, many

prominent scientists found it hardly worth the price. "Not one of my scientific friends thinks it's worth a graduate student's time," one Nobel Prize-winning physicist once told a reporter. "The more-or-less Nobel Prize-winning type of mind thought NASA's program a little too Buck Rogers-ish," a NASA official suggested by way of explanation. It was true that Apollo was primarily a project for engineers, demanding the design and construction of new machines and instruments, not the formulation of new principles likely to attract the attention of the Nobel committee. Moreover, the scientific prizes to be won by exploration of the moon were not in the beginning considered enticing enough to men who were concerned with the mysteries of more distant celestial bodies. This attitude began to change later, however, as the landing drew closer and scientists grew more curious about what the moon might tell them of earth's formation and the history of the solar system.

Typical arguments by scientists against Apollo involved two issues. One was the long-standing debate over the value of manned flight as opposed to unmanned instrumented probes. While conceding that it was too late to eliminate Apollo, Ralph Lapp, a physicist who worked on the Man-hattan project for the atomic bomb, said that it would be more productive scientifically to rely on unmanned instru-mented satellites in the future, and would certainly be less expensive than the "manned spectaculars." Lapp questioned if it was worth all the extra tonnage for life-support equip-ment and the expense for safety mechanisms to put men into space. "We should not forget," Lapp observed, "that the single greatest discovery of the space age, the Van Allen radiation belt, was made with only a 30-pound payload."

The other argument concerned the deadline. After the Pad 34 fire, several scientists and public officials wondered if the haste to meet the goal of a landing in the decade had not contributed to the tragedy. Philip Abelson, editor of *Science*, the weekly journal of the American Association for the Advancement of Science, had previously told a Senate committee that he did not "see anything magical about this decade—the moon has been there a long time and will continue to be there a long time."

Scientists were not alone in their complaints. As early as 1963, Walter Lippmann had raised the same points that were being debated in 1967:

There were two big mistakes. One was the commitment to put a man, a living person rather than instruments, on the moon. The other mistake was to set a deadline—1970—when the man was to land on the moon. These two mistakes have transformed what is an immensely fascinating scientific experiment into a morbid and vulgar stunt. . . . For this is showmanship and not science, and it contaminates the whole affair.

To many Americans the deadline no longer seemed to be critical. Would it really make much difference who got to the moon first? Nine out of ten of those interviewed in a Trendex poll in 1967 said it would not matter.

Some of the same questions must have been raised in the Soviet Union, too. "There's our meat," a young Moscow taxi driver was reported to have said, thrusting his thumb skyward and lamenting the scarcity of food in his country. In what may have been a reply to such complaints, Premier Aleksei Kosygin declared: "Space research expenditures do not affect the needs of the population." Another indication came late in 1966. In a report from Moscow, Radio Prague said that the "difficulty" with both the Soviet and American space programs was that the "issue has become not a scientific but a prestige problem." The report continued: "It costs too much and drains away too many first-class scientific brains, and the two big powers face too many other problems for which both money and brains are required."

On October 4, 1967, the tenth anniversary of the Space Age, NASA's James Webb warned against any retrenchment in American space efforts as a result of criticism stemming from the Apollo fire and the country's preoccupation with other problems. Yet, as he was saying this, the Senate Appropriations Committee approved the lowest NASA budget in five years. While NASA's spending had jumped from $700 million in the year of the decision to $5.9 billion in 1966, the budget was now being cut back to well under $5 billion, with more reductions in sight.

The Case for Apollo

It was not always easy to debate Apollo's critics point for point. The project *was* expensive. There *were* serious problems on earth demanding swift attention. Unmanned probes *were* less expensive and *could* return valuable scientific data. The deadline *was* conceived in the context of the cold war, not for scientific purposes. So conceding, Apollo's defenders sometimes found themselves with nothing firmer to stand on than their sense of history—and their sense of humor.

They liked to quote from a distinguished United States senator's speech: "What do we want of the vast worthless area? This region of deserts, of shifting sands and whirlwinds of dust? To what use could we ever hope to put these deserts or these endless mountain ranges? What use can we have for such a place? I will never vote one cent from the public treasury. . . ." The senator was not referring to the moon; but to California. Daniel Webster, more than 125 years ago, was opposing the appropriation of $50,000 to establish mail service to the Far West.

In a more serious vein, NASA officials insisted that the budget for Apollo was "austere" and "realistic." At $24 billion, they pointed out, it could cost the 200 million men, women and children in America $120 each spread out over nearly nine years. To put this figure in perspective, Lyndon Johnson said, Americans should remember that they spend more on cigarettes, alcohol or recreation. In a single year, he said, the nation "bet more on horse races than on space."

James R. Killian, Jr., who had served as President Eisenhower's first science adviser and was an architect of NASA, also defended Apollo against cutbacks. "Although I regret the circumstances which tend to place and pace Apollo in the context of a race to the moon," Killian said in early 1968, "I feel that at this late stage its funding should not be slowed in a way that would be wasteful of the great investment of money and mobilization of skills we have made in it. A project of this difficulty and magnitude can be gravely hurt by 'on again, off again' support, and in this program we are way past the point of no return."

The in-this-decade commitment, Apollo project officials contended, was more of a help than a handicap. True, it was a factor in all major engineering decisions, such as the selection of one system over another because it could be ready in the prescribed time. But, as one NASA official told me, it "saved us from what we call Brownian motion," which is the phenomenon of the erratic movement of particles in a solution. That is, working with a clearly defined goal (the moon) and deadline (the end of 1969), it was possible to lay out a more orderly schedule of development, production and testing that reduced the chances of wasted motion and thus wasted time and money.

Even though the deadline was a political decision, it was not considered unrealistic. NASA officials denied that it had any bearing on the Apollo 1 fire, although a more leisurely pace might have permitted the incorporation of a less volatile two-gas atmosphere. "The eight years allotted to carrying out the Apollo program," George E. Mueller, NASA's manned spaceflight chief, said, "constitutes a longer period than the duration of any previous United States research and development effort."

As for the either-or argument—if not for Apollo, more money could be spent on poverty, crime, disease or pollution—Apollo's defenders generally found it a valid issue, but one that they suspected was often falsely used. Unquestionably, the Apollo money could be devoted to such worthwhile earthly problems. But those familiar with congressional committees and the Budget Bureau recognized a weakness in the argument. If the space program were ended the next day, there might be a modest tax cut, instead of a tax rise, but would poverty be diminished? Or would it be increased because of the unemployment caused by aerospace shutdowns? Also, some of the public officials using this attack on space were never particularly warm supporters of social welfare endeavors.

Along the same lines, Mueller took issue with the scientists who said that their disciplines were suffering monetarily because of the priority given to space programs. "This philosophy that you chop this to give to that is self-defeating," Mueller maintained. "They have to justify other programs on their own merits. Besides, space programs have brought more money and capability to other scientific fields than any other program we've got."

Turning from defense to advocacy, Apollo's proponents discussed the real raison d'être of the project in broader terms. Man's going to the moon, they said, sprang from scientific, political and technological considerations, and even deeper, from the very roots of the human spirit.

Despite some complaints from scientists, the desire for scientific knowledge was an important factor. Preserved on the surface of the moon—like pages in a book, in the phrase of NASA physicist Robert Jastrow—is a pristine record of the moon's history that could yield precious information about the origins of the solar system. The moon's very barren changelessness may make the moon a museum piece of what worlds were like billions of years ago. The early astronaut-explorers could bring back samples for analysis at leading laboratories around the world. Robot instruments could make certain simple observations, but only men could deal with the unexpected opportunities that might be encountered.

National security was another reason for Apollo. After he made the Apollo decision, President Kennedy often referred to space as the "new ocean," and stressed the need for the United States to occupy a position of preeminence there. Even the critics rarely took issue with this point. Space was clearly connected with two developments vital to defense: one was instruments capable of finding and reaching a distant target, and the other was rocket power. Apollo demanded the most precise instruments and the most powerful rockets.

A third, and related, reason was national prestige. It was probably the main justification in Kennedy's mind. In this view, the race to the moon was a test of national vitality, a demonstration of the nation's ability to lead in an age of technology. Perhaps it would not have made too much difference if the Russians reached the moon first, but it might, indeed, have made a difference if the United States had never tried and was, therefore, not even a close second. Great Britain has failed to keep pace with modern technology, causing her to slip well down in the second rank of world powers and to suffer economically at home and abroad. In the mind of Kennedy, where the project got its real mandate, space exploration was "one of the great adventures of our time, and no nation which expects to be the leader of other nations can expect to stay behind."

Still another factor was technological. Flying men to the moon *was* possible. In 1961, after due deliberation, it was found that the United States could muster the rockets, computers, radar, tracking antennas and communications equipment necessary for such a venture. And when man learns he can do something new, it is seldom long before he gives it a try.

Beyond that was the consideration that a massive project of the scale of Apollo, with a single focus, would more than anything else mobilize the nation to meet the challenges and seize the opportunities of the Space Age. It would draw upon existing ("state-of-the-art," as space engineers call it) technology and also, because of its unusual demands, generate a whole new body of technology. In this way it could do what wars had done in the past: for example, World War I spurred development of the airplane and World War II led to advances in radar, electronics and the harnessing of atomic energy. For a time some optimists even talked of space competition among nations as the moral equivalent of war. But the widespread violence of the decade indicated that they were premature, at best.

Apollo, therefore, was not just a project to land two men on the moon. "When Charles Lindbergh made his famous first flight to Paris," wrote Wernher von Braun in 1967, "I do not think that anyone believed that his sole purpose in going was simply to get to Paris. His purpose was to demonstrate the feasibility of transoceanic air travel. He had the farsightedness to realize that the best way to demonstrate his point to the world was to select a target familiar to everyone. In the Apollo program, the moon is our Paris."

It might be years before the full technological import of Apollo was realized. But NASA could already point to some "spin-offs." Apollo engineers had to design valves, pumps, filters and switches that function with a reliability never before achieved. They created miniature medical sensors of unprecedented sensitivity to monitor the astronauts' reactions to stress. They vastly extended the range of communications equipment. They developed compact new electronic parts, new long-life power sources, new alloys, new adhesives, new lubricants. They invented ingenious tools and techniques for shaping and joining metals.

But underlying the specific scientific, military and technological reasons for exploration of the moon lay a deeper, more universal motivation. This was man's insatiable curiosity, the undying restlessness that had sent explorers to brave the harshest conditions on this planet—to the poles, to the tops of the highest mountains, to the loftiest balloon altitudes and to the deepest portions of the seas. Then why not to the moon?

This commingling of the quest for knowledge with a love of adventure often can motivate the man in the street as well as the scientist or explorer. His participation might be vicarious, as with a television watcher captivated by the make-believe of science fiction or the reality of a mission to the moon. In any event, the lure of adventure is not to be sneered at, not if it lifts man's sights higher. Who can say what would be the ultimate worth of such a grand endeavor as Apollo?

Martin Schwarzschild, a Princeton University astronomer, once gave the project's defenders an eloquent argument. "We must not justify going to the moon as a single project," Schwarzschild said. "It must be justified [in terms of] the spirit of a movement, under the wing of which the really valuable things must happen."

From the Pyramids, he pointed out, it was the bust of Nefertiti that was of lasting value, and it might never have been done if the Pyramids had not been built. From St. Peter's Basilica, which Schwarzschild found architecturally distasteful, it was Michelangelo's *Pietà* that was worth the price of the whole project. "Whenever there is real energy in something," Schwarzschild commented, "it stimulates the energy in other fields—it does not reduce it." Historians have made the same point about the Renaissance and the fact that it may be no accident that this era of cultural advance took place at the same time as the Age of Exploration, the time of Columbus, Cabot and Magellan.

The Firm Commitment

Even the critics, in 1967, seemed willing to give Apollo the benefit of the doubt. They would let it continue its course, if it could, through its Sargasso Sea and toward its destination. In many cases, they were more intent on influencing federal spending beyond Apollo, tempering any impulse to select a new goal such as Mars, reminding mankind of its unachieved adventures on earth. Though the overall national space budget was reduced, the specific funds for Apollo were left relatively untouched.

It is a testimony to President Kennedy's understanding of the American competitive spirit that he made the Apollo decision. He liked to tell a story about the importance of commitment, a story that Apollo advocates often repeated. It was a story by the Irish-American writer Frank O'Connor.

"As a boy," Kennedy would begin, "O'Connor and his friends would make their way across the countryside, and when they came to an orchard wall that seemed too high and too doubtful to try and too difficult to permit their voyage to continue, they took off their hats and tossed them over the wall—and then they had no choice but to follow." Kennedy would pause for effect. "This nation has tossed its cap over the wall of space, and we have no choice but to follow it."

And so Apollo, after the nation's cap, moved slowly and carefully toward its first major test since the fire. Asked what would be needed to overcome the doubts and criticism, James Webb told a reporter: "Nothing, until we make a couple of these birds fly."

The bird, Saturn 5, was on the pad at Cape Kennedy and ready for its first flight test in November. Much was at stake.

Chapter X

The Moon Rocket

Twenty-five years after he fired the first successful German V-2 rocket and almost ten years after launching the first American satellite, Wernher von Braun had harnessed enough power in one rocket to send men to the moon. It was by far the largest rocket ever built—nearly 200 times the weight of the V-2 and four times more powerful than any American or Russian rocket ever launched. It was the Saturn 5, "the big bird," and its first trial by flight was scheduled for November 9, 1967, at Cape Kennedy.

I took the long way to the Cape, stopping off at Huntsville, Alabama, to visit with von Braun and a number of the Saturn 5 engineers at the Marshall Space Flight Center. For von Braun, then 55 years old, whose long, dark blond hair was turning gray, the Saturn 5 represented a culmination of rocket experimentation he began as a college boy in Berlin. It also marked the introduction of a vehicle capable of initiating the interplanetary travel which he had decided as a young man was a task "worth dedicating one's life to." "For my confirmation," said von Braun, "I didn't get a watch and my first pair of long pants, like most Lutheran boys. I got a telescope."

At the age of 20, after helping launch dozens of primitive rockets from a Berlin field, von Braun became the top civilian specialist for the German army's new rocket station at Kummersdorf. That was in 1932, and it led to his involvement in the development of Hitler's V-2. When the first V-2 blasted London, von Braun, the space enthusiast, remarked to a friend that the rocket had worked perfectly except for landing on the wrong planet.

After 22 years in the United States, von Braun still spoke with a distinct accent. One arm was always shooting out to illustrate a rocket's curving course or to identify some planetary target in midair. He was given to analogies between space exploration and Charles Lindbergh, the Wright brothers and Christopher Columbus. When he completed an analogy he liked particularly well, he would break into a slightly lopsided smile, arch his eyebrows and look to make sure the visitor got his point. On a long cabinet behind his desk stood 11 scale models of rockets, stair-stepping up from the Pershing missile and the V–2, about a foot high, to the Saturn 5. Even as a model it towered so high above earlier rockets that von Braun had to have a hole cut in his ceiling for the Saturn 5 model to stand upright.

"You shouldn't just consider it an overgrown V–2," von Braun told me. "That's like saying the Boeing 707 is only an overgrown Wright brothers' plane. About the only thing the V–2 and the Saturn 5 have in common is that they both operate on Newton's third law that every action has an equal and opposite reaction."[*]

Von Braun's considerable contributions to Apollo extended beyond the technology of rocketry. Far more than any other engineer or manager in the project, even more than most of the astronauts, von Braun had charisma, that indefinable ability to inspire men, and a sense of the larger purpose for which men ventured into space. He was NASA's biggest star, and certainly the only one who had a Hollywood movie (*I Aim for the Stars*) made of his life and dreams for interplanetary exploration.

Anatomy of Saturn 5

Everything about the Saturn 5 was big. When topped with the Apollo spacecraft, the three-stage rocket stood 363 feet tall, six stories higher than the Statue of Liberty. Its first, or booster, stage was the biggest aluminum cylinder ever machined. Its valves were as big as barrels, its fuel pumps (for feeding engines at the rate of 700 tons of fuel a minute) were bigger than refrigerators, its pipes were big enough for a man to crawl through and its engines were the size of trucks.

---

[*] Although the largest and most advanced missile of its time, the V–2 was only 46 feet long, 65 inches in diameter and weighed 27,000 pounds. It had only one stage and one engine, which burned alcohol mixed with liquid oxygen and generated a maximum thrust of 56,000 pounds. By contrast, the three-stage Saturn 5 is over 280 feet long, 33 feet in diameter and generates a lift-off thrust alone of 7.5 million pounds.

While Saturn 5 followed a series of smaller rockets, such as Redstone and Jupiter, its development posed many new challenges for the von Braun team. There were difficulties in finding the right materials and propellants, in achieving successful welding and assembly techniques. And because of the rocket's huge size, there were new problems in creating methods for transporting and launching the Saturn.

The Saturn 5 came in four main sections:

First Stage (S–I).—The Saturn engineers called it the "most mature" of the rocket's units because it was basically an enlargement of previously tested propulsion systems. Those who criticized von Braun as being too conservative an engineer contended that the stage's kerosene and liquid oxygen fuel mixture was obsolete, only a slight advance over the V–2.

By itself the stage, built by the Boeing Company, was bigger than any single previous rocket—138 feet tall and 33 feet in diameter. It weighed 307,000 pounds without propellants. With propellants (203,000 gallons of refined kerosene, RP–1, and 331,000 gallons of liquid oxygen in two separate tanks), the weight jumped to 4,707,000 pounds, more than three-quarters the weight of the entire Saturn 5 and Apollo vehicle.

Clustered at the base of the stage were five engines, each 18 feet tall, that consumed all that fuel in 2½ minutes to force a fiery exhaust out their five bell-shaped nozzles. Someone figured it out that the five engines, built by the Rocketdyne division of North American Rockwell, burned oxygen at the same rate it is consumed by half a billion people and fuel at the rate it is used by 3 million automobiles running simultaneously. Each engine (designated the F–1) generated 1.5 million pounds of thrust, making the total thrust of the stage 7.5 million pounds. With such power it would be possible to boost the 6.2 million pound Saturn-Apollo machine 38 miles through the atmosphere to a terminal speed of almost 6,100 miles an hour. At that point the spent stage would disconnect and drop away.

Second Stage (S–II).—Perched on the shoulders of the booster was the second stage, the most powerful hydrogen-fueled vehicle ever built. The entire stage—81 feet 7 inches high and 33 feet wide—was essentially two thin-walled

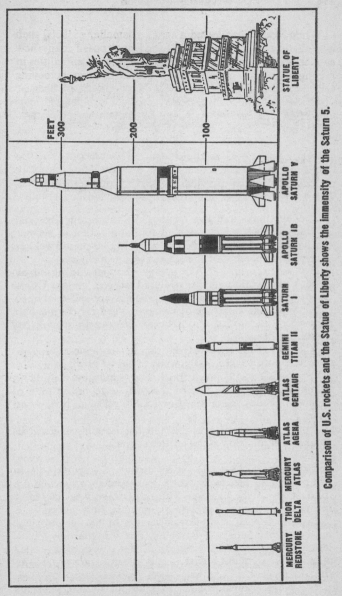

FEET

300

200

100

MERCURY REDSTONE — THOR DELTA — MERCURY ATLAS — ATLAS AGENA — ATLAS CENTAUR — GEMINI TITAN II — SATURN I — APOLLO SATURN IB — APOLLO SATURN V — STATUE OF LIBERTY

Comparison of U.S. rockets and the Statue of Liberty shows the immensity of the Saturn 5.

The Saturn 5—its major features, shown with other components of an Apollo space vehicle.

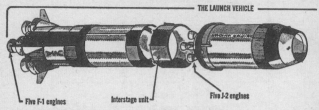

Five F-1 engines   Interstage unit   Five J-2 engines

Five F-1 engines: Launch vehicle with 7.5 million pounds of thrust. **Interstage unit:** Connects first two stages and contains equipment for both. **Five J-2 engines:** Take rocket to altitude for earth orbit with more than a million pounds of thrust. **Single J-2 engine:** Puts vehicle into orbit with 200,000 pounds of thrust.

propellant tanks and five engines. By machining the walls as thin as possible and having a common bulkhead between the tanks, engineers arrived at a 1,033,000-pound vehicle, in which 90 percent of the weight was propellants.

The tanks held 267,700 gallons of supercold liquid hydrogen and 87,000 gallons of liquid oxygen, or lox. Unlike air-breathing jet engines, rocket engines are self-contained. They must take along the necessary oxygen for combustion; otherwise, in rarefied atmospheres and out in space, they could not fire at all.

The five J–2 engines of the second stage were designed to develop up to 225,000 pounds of thrust each, or 1,125,-000 pounds for the entire stage. They could burn some six minutes, pushing the rest of the vehicle to an altitude of about 115 miles and a speed of 15,300 miles an hour—just short of orbit.

North American Rockwell built the second stage and its Rocketdyne division supplied the engines.

Third Stage (S–IVB).—The Saturn 5 tapered off in width from 33 feet to 21 feet 8 inches when it reached the third stage of the "stack." The third stage was 58 feet 7 inches high and weighed 260,000 pounds, fully fueled.

Because it was an improved version of the second stage on the earlier Saturn 1–B rocket, this was the only flight-tested unit to go into the Saturn 5. But it was being called upon to do something that it had never done before—restart in space.

The stage's single engine, identical to the second-stage J–2's, was supposed to fire when the vehicle was approach-

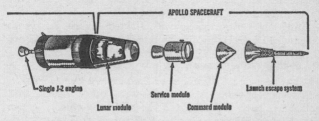

Lunar module: On lunar flight, will take two astronauts to the surface of the moon. Service module: Contains support systems for command module and power to maneuver it. Command module: "Home" for the three astronauts on their way to the moon and back. Launch escape system: Carries command module free of rocket in case of accident.

ing orbit. By firing more than two minutes, boosting its speed to 17,400 miles an hour, the third stage would push itself and the attached Apollo payload into earth orbit. On a voyage to the moon, the engine would be required to reignite for about five minutes to boost the astronauts out of their low earth orbit and on the way to the moon.

The third stage carried 63,000 gallons of liquid hydrogen and 20,000 gallons of liquid oxygen. It was built by the McDonnell Douglas Corporation.

Instrument Unit.—This three-foot-high section housed the rocket's electronic brain, Designed and built by the International Business Machines Corporation, the 4,750-pound unit contained the electronics and electrical equipment, including a computer, needed for the guidance and navigation of the rocket's flight. Signals from the instrument unit pulsed through the rocket to ignite and shut down the engines and monitor their performance.

Putting It Together

The Saturn 5's very bigness was the source of most of the problems encountered during its development and construction. "It was all harder than I ever expected," Arthur Rudolph, the project manager, said when I talked with him at Huntsville. "Since it is so large and has so many components, you get more people involved, and that makes it difficult. You have to get them all to sing from the same sheet of music. My main task has been as choir director."

Like most of the key engineers running the project, the bald, pink-faced Rudolph was a Peenemünde veteran. He first joined forces with von Braun in 1934 and remembered igniting his early rockets with lighted cigarettes. Of the original 118 German engineers who came to the United States with von Braun, 64 were still working with him at Huntsville. Thirty were working elsewhere in the country; 15 had returned to Europe, and nine were dead. Working with Rudolph—in his "choir"—were other Peenemünde colleagues such as William A. Mrazek, the chief engineer, and Walter Häussermann and Ernst D. Geissler. Some 100,000 people and several thousand industrial contractors also had a hand in designing, building and testing the Saturn 5 at a cost of $7 billion for the 15 rockets on order.

The Huntsville engineers were instructed to start work on the moon rocket in January, 1962, using F–1 kerosene engines that had been under development since 1958 and the hydrogen-fueled J–2 engines, upon which work had begun in 1960. Although this advance engine research proved a help, the biggest headaches lay ahead. Von Braun, Rudolph and the other engineers had to contend with many frustrations before getting the higher-strength alloys they needed for the rocket walls and finding ways to weld the mammoth parts together. Of some 3,500 components considered at the start, about 1,000 did not meet specifications and had to be redesigned. Making compromises between what was wanted and what was possible took much of Rudolph's time. "You want a valve that doesn't leak and you try everything possible to develop one," Rudolph said. "But the real world provides you with a leaky valve. You must determine how much leaking you can tolerate."

Putting the big machine together turned out to be one of the toughest manufacturing problems. No one had ever attempted to weld pieces of ultrathin aluminum alloy for a cylinder 33 feet in diameter. A weak weld could cause the whole vehicle to blow to pieces. After months of experimentation, the contractors developed automatic weld machines that moved along a precision track. The seams were then inspected by X ray and reinforced, where weak, with hand-controlled welding. The first stage alone required nearly 8,000 feet of welding seams.

The decision to use liquid hydrogen as the fuel for the second- and third-stage engines presented the engineers with another challenge to their ingenuity. It was also another case of the Saturn 5 builders following the suggestions made by the farsighted Konstantin Tsiolkovski, the Russian astronautics pioneer, at the turn of the century. He had proposed building rockets by stages, which could be discarded after use to reduce the total weight. And he had observed that liquid hydrogen would make a more efficient rocket fuel.

Until the end of World War II, however, liquid hydrogen could not even be manufactured and stored in quantities large enough to run a rocket. The problem: hydrogen does not want to be a liquid. To stay liquid, it must be kept at −423° Fahrenheit; otherwise it reverts to a gas and evaporates. This meant it had to be stored in heavily insulated tanks that were all but leakproof. Experience with the Centaur, a smaller hydrogen-fueled rocket under development in the early 1960s, was anything but encouraging. One exploded after lift-off in May, 1962, when an insulating panel ripped off too soon, exposing the hydrogen to heat.

For the Saturn 5's third stage, the contractor spent two years finding plastic insulation for the tanks, pipes and valves, for everything that came in contact with the supercold propellants. (Liquid hydrogen burned on contact with liquid oxygen, which had to be kept at a temperature of −293°.) The second-stage contractor had an even tougher time of it because the tank walls doubled as the walls of the rocket itself. A glass fiber honeycomb was coated on the outer surface to make the stage like a giant thermos bottle.

It was worth the trouble. Liquid hydrogen is the lightest of elements. One gallon weighs six-tenths of a pound, compared to eight pounds for the same amount of kerosene. This meant that more liquid hydrogen could be carried in a smaller tank. Moreover, liquid hydrogen's light and simple atoms could be accelerated out the rocket nozzle at a higher speed than anything else, causing a more powerful forward reaction. All of this added up to more miles to the gallon from liquid hydrogen.

The Saturn team had worked toward an initial launching test of their big rocket by 1966 or early 1967. But the manufacturing problems and two explosions on the test stand delayed the final readiness. Then, when they were finally ready, the engineers were sure of their great bird. "We have no doubts at all that the Saturn 5 should be able to make it," von Braun told me. "It's not like the early days. Then it was a kind of trial-and-error method. You designed these things but didn't really know what they would do until you pushed the button."

Through the use of computer analysis, vibration tests, simulations of the space environment and ground test firings, von Braun explained, it had become possible to weed out or redesign potentially troublesome parts in the rocket. "We have really tortured all the systems and components against all conceivable environmental conditions the rocket might see during launch or flight," von Braun said. "And we have an automated electronic check-out system that monitors thousands of points on the rocket right up to the moment of lift-off so that the probability of launching a sick system is quite remote."

Moonport

Just to get the many parts of the Saturn 5 to Cape Kennedy was no mean feat. The first stage went by special barge from the Mississippi Test Facility on the Gulf of Mexico. The second stage was shipped by barge from California through the Panama Canal. The third stage and other smaller components were sent to the Cape in an oversized transport plane called the Super Guppy, so named because it looked like a bloated fish. The three stages were assembled—"mated," as the engineers called it—in a 525-foot-tall building near the launching pad. Thus, besides being the first test of the moon rocket, the Saturn 5 launching would also be the moonport's inaugural.

In many ways, the buildings, fuel tanks, launching towers and control rooms of the moonport were as complex as the Saturn 5 itself. They were all built to a scale matching the Saturn 5's. And they also presented engineers with a number of difficult construction problems.

The man in direct charge of the moonport was Rocco A. Petrone, a former air-force lieutenant colonel and West Point football player whose broad shoulders seemed big enough for the problems and whose quick mind seemed to sort and file the smallest details of the mammoth undertaking. One day I went on a tour up to the top of the launching-pad tower and up to the top of the Vehicle Assembly Building (called the VAB) where the rockets were being stacked together. Afterward, while stopping by Petrone's office and looking out his window toward the pad, I believe for the first time I saw that huge mound of reinforced concrete as the future national monument it will surely become—a place tourists and schoolchildren will be ushered through while guides explain that, back in 1969, from this very spot, men took off in a rocket and went and landed on the moon. But Petrone, the engineer and manager, was talking of the problems of getting the facilities ready for its first of many tests. "Concurrency is the real challenge of this program," he was saying—getting everything ready at the same time.

Until Boeing put up its sprawling jet-aircraft plant in Seattle, the Vehicle Assembly Building at the moonport contained the largest volume of any edifice in the world—129,482,000 cubic feet. Its enormous dimensions: 525 feet high, 716 feet long and 518 feet wide. But from the top of the launching tower, 3½ miles away, it did not seem overwhelming, like the Empire State Building, because there were no dwarfed skyscrapers nearby. Then a sharpened eye detected specks moving about—trucks, cars and people—and the building came into better perspective.

Extending diagonally from the southeast corner of the building was a long four-story structure—the launch control center. It was the current substitute for the old-fashioned blockhouse that had to be much closer to its launching pad because the electronics experts had not yet developed launching equipment that could be operated from very far away. Unseen beneath the ground were the miles of cable that spewed billions of bits of data back and forth between the control center and the rocket on the pad.

The VAB was designed by a combine of four New York engineering-architectural firms. Max O. Urbahn, the managing partner of the group, had discussed the project when it was just beginning:

> As an architectural structure, the building will be little more than a slick polished box, covering eight acres. Inside the box it is somewhat more interesting, with whole buildings hanging from its sides, some of them

moving up and down and in and out like suspended file drawers, mating with one another to form still other buildings within buildings to house the space vehicles.

We have solved some intriguing technical problems created by the monstrous size. For instance, more than 50,000 tons of structural steel will be needed in the VAB's framework, enough for more than 30,000 automobiles.

We were faced with the fascinating possibility that the shape of the building might make it react like an immense box kite, and it could blow away in a high wind. We had to design pile foundations that would prevent that.

The area's sandy soil was another reason for needing deep pile foundations for the building. Workers had to penetrate 160 feet to reach solid limestone bedrock in which to drive steel roots—4,225 of them—to support the structure. Because of the building's size it was necessary to install an air-conditioning and ventilation system to prevent the formation of clouds inside such a vast area.

Inside the VAB the Saturn 5s took shape, stage by stage. They were erected vertically on top of a two-story, steel-plated base with a 398-foot-high tower. This was called the mobile launcher, and steel arms would hold the rocket in place through all the tests and up to seconds after ignition.

Once the rocket was assembled, it had to be moved out to the pad. Proposals to move it by barge or by rail were considered and rejected; a barge seemed to be too unstable, and the weight of the rocket might flatten steel railroad wheels. In the end, someone recalled the steam-shovel tractors used in the open-pit coal mines of Kentucky. This led to the development of the crawler-transporter, a squat vehicle driven by eight tanklike tractors, two at each corner. The vehicle was as big as a baseball diamond.

A special roadway, packed with river rock, was laid between the VAB and the two Saturn 5 launching pads—39–A and 39–B. The rocket's 3½ mile journey by crawler-transporter took nearly eight hours—the slowest part of man's trip to the moon.

At the launching pad another steel-trussed tower was provided to enable technicians to get to the rocket to check it out and fuel it. Called the mobile service structure, the 40-story tower had five platforms that completely enclosed the rocket. The structure was hauled away 11 hours before lift-off, leaving the Saturn 5 and its support tower alone on the pad.

Apollo 4

This was the way the first Saturn 5 was before the dawn on the morning of November 9, waiting for its maiden flight: enveloped in a flood of xenon lights, standing out against the dark sky like a crystalline obelisk. As I approached it, driving from Cocoa Beach toward the press site, the rocket loomed like some strange nocturnal mirage, a bright specter that would no doubt vanish with the morning dew. I recalled a conversation two days before at Ramon's, the missilemen's favorite eating and drinking spot on Cocoa Beach. B. B. MacNabb, who had worked for more than a decade as operations manager for 109 Atlas launchings, had come from Valley Forge on his vacation to see the Saturn 5. He was filled with nostalgia for the old days and excitement over the launch to come. Said MacNabb: "You'd have to be dead from the neck up—or down—not to get excited about this."

If the sight of the Saturn 5 was breathtaking, the sound and vibrations of its launching brought you back to reality. As the 49-hour countdown ticked toward its final three minutes, the rocket went "automatic." Its own instrument unit ran the last checks and then triggered the five engines of the first stage to fire in sequence, sending gushes of orange flame and smoke over the launching pad. A million gallons of water rushed into the concrete "flame bucket" underneath to prevent serious damage to the pad. Once it was automatically determined that all five engines were firing evenly, the steel "hold arms" on the mobile tower withdrew their restraining grip. At 7 A.M., on schedule, the 36-story vehicle began to rise from its pad and its mushroom of smoke.

"Go, baby, go," von Braun shouted as he watched anxiously through binoculars.

Not until the rocket cleared the launch tower, about ten seconds after lift-off, did the shock wave reach the press site three miles away. It traveled through the ground and the air. Staccato bursts, rapid and thunderous, shook the small wooden grandstand, rattled its corrugated iron roof and pushed in the plate-glass window on the Columbia Broadcasting System's mobile studio. Reporters in the open could feel the pressure of the shock wave beating at their faces and chests. Inside the launch control center, the consoles were covered with plaster dust that had shaken loose from the ceiling.

Each of the Saturn's three stages fired its allotted time. When the third stage shut down, 11 minutes after lift-off, the unmanned Apollo spacecraft that had been on top of the rocket was orbiting earth. Unlike the other stages, the third remained attached to the spacecraft. Later, it refired to drive the spacecraft 11,286 miles out from earth.

The rocket's third stage and a dummy Lunar Module, taken along to make it a realistic test of the full Saturn-Apollo "stack," were then jettisoned. On a signal from ground controllers, the Apollo's main rocket, housed in the Service Module, fired to ram the Command Module back through the atmosphere at 24,500 miles an hour, the speed it would travel on a return from the moon. This gave the Command Module's heat shield its first test at such reentry speeds and temperatures. After a mission of eight hours and 40 minutes, the Command Module splashed down in the Pacific northwest of Hawaii.

All elements of the mission, designated Apollo 4,* passed their tests. The launching complex, the Saturn 5 and the Command Module all were reported to have performed without flaw.

"That was the best birthday candle I ever had," exclaimed Arthur Rudolph, who turned 60 years old on launching day.† A fresh breeze of optimism stirred through Cape Kennedy after the mission, dispelling many of the doubts and much of the gloom that had filled the place since the launching-pad fire ten months earlier. At one of the "splashdown parties," an engineer for the widely criticized North American Rockwell was heard to proclaim:

---

* Apollos 1, 2 and 3 were the unmanned missions to check out the Command Module and the S–IVB rocket stage. All three were launched by Saturn 1–B rockets in 1966. The numbers were applied to the missions ex post facto. This led to some confusion because most accounts of the fire that killed Grissom, White and Chaffee referred to their planned flight as Apollo 1.

† Rudolph retired in 1968 and was succeeded by Lee B. James, a retired army colonel who had worked on guided missiles and space vehicles since 1947.

"Our slate's clean now, baby! We showed them who knows how to fly a bird."

Apollo 6

The second flight of the Saturn 5 was, however, less than a full success. On April 4, 1968, another of the rockets was launched in a test with an Apollo spacecraft—the Apollo 6 mission. (Apollo 5, in January, was the first unmanned test of the Lunar Module and went up with a Saturn 1B.) But two of the second-stage engines shut down prematurely and the third-stage engine failed to reignite in orbit.

For a time it appeared that NASA would have to conduct a third unmanned Saturn 5 launching before the mighty rocket could be "man-rated"—cleared for launching men. But engineers at Huntsville, after weeks of intensive detective work, determined that the engine cutoffs were caused by extreme vibrations and that these could be damped out.

The Saturn 5 was thus declared ready for its role in sending men to the moon. And the United States, for the first time in the Space Age, had a rocket that far surpassed the lifting capabilities of anything the Soviet Union had tested.

Chapter XI

The Landing Bug

The theory that a form dictated by function, unadorned, is beautiful was put to a severe test by the Lunar Module, the spidery-looking landing vehicle. Although a century from now it may stand, in the manner of the automobile classics, as a reminder of an age when men still knew how to design a good-looking spacecraft, Tom Buckley of *The New York Times* Sunday magazine wrote that the Lunar Module of the 1960s "looks like an earthling's nightmare of a visitor from another planet." In an article published in February, 1969, Buckley gave the following description:

> An ungainly, underslung octagonal base is supported by four spindly, awkwardly jointed legs. The base is surmounted by a faintly cranial-looking structure—essentially a stubby cylinder, but one that has been hopelessly deformed by a number of eccentrically shaped and sharply angled protuberances—from which a hatch gapes like an idiot's mouth and two triangular windows glare like eyes. Four bathtub handles, the skirts of the small rocket engines that control the LM's attitude in flight, jut from the cylinder's sides, and its top is garnished by a collection of antennas—dish, curled and ball-topped—that look like flowers on an old woman's hat. The LM manages to suggest simultaneously a not-very-well-thought-out cubist sculpture, the machine sculpture of Jean Tinguely and the Edsel automobile.

Was this any way to talk about the vehicle entrusted with the crucial and climactic task of ferrying the two astronauts down to the moon and back to the lunar-orbiting Command Module? Perhaps not, but there was no way of looking at the Lunar Module without head-nodding amusement and, eventually, a feeling that that funny-looking squat body with its legs and antennas was somehow animate.

Engineers who worked on the Lunar Module usually called it the LM—pronounced "lem," which was a carry-over from its original name of Lunar Excursion Module. (The "excursion" was dropped by NASA because of what some people regarded as a frivolous connotation.) Nearly everyone besides the engineers applied such appellations as the spider, the moonbug, the landing bug. For compared with the massive but streamlined Saturn 5 and the smooth, cone-shaped Command Module, the Lunar Module was the ugly duckling of the Apollo vehicles.

In defense of his bizarre creation, Joseph G. Gavin, Jr., vice president of space programs for the Grumman Aircraft Engineering Corporation, the prime contractor, reminded visiting reporters that few airplanes really achieved grace but, rather, had it thrust upon them to reduce wind resistance and improve stability in flight. The shape and structure of the Saturn 5 was dictated by the fact that it must fight its way upward through the thick earth atmosphere. The Command Module looked the way it did because it must withstand the stresses and intense heat of reentry into earth's atmosphere. But the LM would operate only in space, primarily on and around the airless, low-gravity moon. During the launching phase, it would ride in a protective aluminum shell between the Saturn 5's third stage and the Service Module, the rear equipment unit of the Command Module. Unlike the Command Module, the LM would never return to earth after its job was done, thus eliminating the need for any streamlining or shielding against atmospheric friction.

The appearance of the LM was, in any event, the least of NASA's and Grumman's concerns. The LM was the last of the Apollo "hardware" to be developed, its start having been delayed while NASA made up its mind to take the lunar-orbit rendezvous approach—rather than two other possible approaches—and thus require a vehicle like the LM for a landing. Grumman's contract as the prime builder was not signed until January, 1963, almost two years after the Apollo project began.

Another handicap was the lack of prior experience. While the Saturn 5 engineers could draw on the knowledge gained from decades of rocket building, and the Command Module designers had the Mercury and Gemini space capsules as guides, no American spacecraft had been built to do anything similar to what the Lunar Module was supposed to do. The unmanned Surveyors did not fly their lunar-landing missions until 1966. No other spacecraft was being contemplated to not only land on a solid surface, such as the moon, but also to take off again under its own power.

It was, therefore, not surprising that the Lunar Module ran into many difficulties and delays during its development and eventually cost much more than had been anticipated: at least $1.6 billion instead of the original estimates of $500 million to $800 million. This led Joe Gavin, a slender, graying, soft-spoken Bostonian with deep-set eyes, to remark: "A lot of people may look at the Lunar Module and say to themselves 'if I did it myself in the cellar it would be a snap.' But they forget that every piece of material must have a pedigree, that the tools must be super clean and, above all, that there would be no instruction sheet. We had to figure it out for ourselves."

How well Gavin and Grumman, NASA and thousands of other engineers figured out how to build the nation's lunar landing craft was first put to a flight test in January, 1968.

LM's Two Stages

The first flight model of the Lunar Module was a far cry from the landing craft envisioned in the earliest concepts. When the lunar-orbit rendezvous method was being considered, three basic LM models were considered by Langley Research Center engineers. While some of the proposals differed radically, the three main concepts were the shoestring, the economy and the plush.

The shoestring version was hardly more than an open-top Dick Tracy space coupe designed to contain one man in a spacesuit for several hours. It would weigh as little as two tons. The larger economy model, designed to carry two men, would weigh two or three times as much as the shoestring, depending on the types of propellants used. The plush model was eventually decided upon as the safest. By the time it emerged from the Grumman drawing boards it was a "super" plush model with so many additions that it was forever having a weight problem.

Grumman was a company based on New York's Long Island, at Bethpage, that specialized in military aircraft, particularly for the navy. After losing out on bids for the Mercury contract, Grumman engineers, under Gavin, turned in 1960 to examining the types of vehicles for a moon landing. Their foresight paid off when they were able to present NASA with a mountain of predigested data on lunar-landing equipment and thereby win out over eight other aerospace companies for the Lunar Module contract.

"During the original proposal stage," Thomas J. Kelly, the LM's principal design engineer, recalled, "we were thinking in terms of 5,000 pounds and things just grew from there." Kelly, a short, dapper Irishman, was the man at Grumman most deserving of the title "Mr. LM."

To Kelly the LM was 12 tons of propellants surrounded by four tons of "watchmaker's structure" encased in an eggshell-thin aluminum fuselage. It stood 22 feet 11 inches tall and had a diameter of 31 feet with its four legs extended. It had a million parts, most of them tiny transistors, 40 miles of wiring, two radios, two radar sets, six engines, a computer and a lunar scientific experiment kit. All this came in two main parts, known as the ascent and descent stages, each with its own rocket:

Descent Stage.—This was the unmanned lower part of the LM. Made of aluminum alloy, the octagonal-shaped stage with the four legs for landing contained the batteries, oxygen supply and scientific equipment to be used for the descent to the moon and during the astronauts' stay there. The stage stood 10 feet 7 inches tall, counting the legs, and had a diameter of 14 feet 1 inch, not counting the extended legs.

Outriggers extending from the end of each of the two main beams of the stage provided the supports for the landing gear. The legs were extendable once the LM was deployed in space. Each leg consisted of a primary strut, two support struts, locking mechanisms and a footpad. All the struts had crushable shock-absorbing honeycomb inserts to cushion the landing. The forward landing gear, extending below the front hatch, had an attached ladder on which the astronauts would climb down from the ascent stage to the surface and back again.

Most of the descent stage's weight and space were devoted to the four propellant tanks and the throttlable

10,000-pound-thrust descent rocket. Built by TRW, Inc., the engine was rare in rocketry in that, through computer and hand controls, it could be throttled down to 10 percent of its power or up to 94 percent. The engine burned a liquid fuel of hydrazine and unsymmetrical dimethyl-hydrazine, known as aerozine 50, and nitrogen tetroxide as the oxidizer. The fuel and oxidizer burned on contact without the need of a spark.

On the lunar-landing mission, the descent engine would be fired to begin the LM's drop from 70 miles out in lunar orbit down toward the moon. At an altitude of about 50,000 feet, the engine would be refired in another braking maneuver so that the LM would drop slowly down to a soft touchdown on the surface. Because the LM could descend vertically and hover above the surface, the astronauts, who were qualified on jet aircraft, had to qualify also as helicopter pilots. For the final drop the astronauts were supposed to ease back on the power and let the LM fall at a speed of about three feet a second. Five-foot wands, like curb-feelers, would automatically extend downward from the LM's footpads and signal the first contact. The astronauts would then cut off the engine.

After the two men finished their stay on the surface, the descent stage would serve as the launching base for the ascent engine's firing to boost the upper half of the LM off the moon. Explosive devices would be triggered to separate the two stages. The descent stage then would remain on the moon as a relic of man's first landing.

Ascent Stage.—This was the roundish upper half of the LM, the command center and crew cabin as well as the launching rocket for leaving the moon. It was 12 feet 4 inches high and divided into three sections: crew compartment, midsection and equipment bay.

The crew compartment, only 92 inches in diameter and 42 inches deep, took up the front end of the ascent stage where the astronauts could look out the two triangular windows. To save weight there were no seats for the men; they would stand, loosely harnessed by straps. In front and on either side of them were the control panels for the LM's guidance, communications, environment and propulsion systems. Above the commander's position, on the left side, was a window through which he could look to

steer the LM back for the rendezvous and docking with
the Command Module. At the astronauts' feet was the
42-inch-square forward hatch through which they would
leave the LM to walk on the moon's surface.

Both the crew compartment and the midsection were
built and insulated to be pressurized with 100 percent
oxygen. The midsection housed many of the environmental,
communications and guidance systems. It also had a place
to store the containers of moon rocks the astronauts would
take back with them. Overhead in the midsection was the
33-inch-diameter hatch where the astronauts transferred

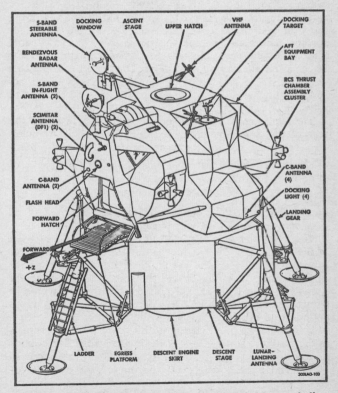

The two-stage Lunar Module, the ungainly "spider" that transports the
astronauts to the moon's surface.

to and from the Command Module when the two vehicles were linked. The LM's overhead hatch would open to a short connecting tunnel between the LM and the nose of the Command Module.

Below the deck of the midsection was the ascent rocket. It was built by North American Rockwell's Rocketdyne division* to generate 3,500 pounds of thrust and to be started and restarted but not throttled. Such a small rocket was sufficient because most of the LM's original weight of 32,000 pounds would have already been reduced to about 10,000 pounds by leaving the spent descent stage on the moon. The moon's weak gravity—one-sixth that of earth's—meant that the LM would not require a great burst of propulsive energy to pull away from the lunar surface. The ascent engine was designed to burn the same kind of liquid propellants as the descent engine and the 16 small maneuvering thrusters. The ascent propellants were stored in two spherical titanium tanks, which accounted for two of the bulges on the side of the ascent stage.

The equipment bay was an unpressurized area in the rear of the stage that contained units to cool the vehicle's electronics, the breathing oxygen for the ascent and a number of other components not requiring pressurization.

Watching the Weight

Most of the LM's troubles during the eight years of development had to do with weight and reliability.

The LM's size was restricted by the diameter and length of the segment of the 363-foot Saturn-Apollo vehicle where it was carried. Its weight was limited by the lifting capacity of the Saturn 5. At first it was thought that the LM could weigh no more than 25,000 pounds, but improvements in Saturn's efficiency gave the LM engineers 7,000 more pounds to work with. Even so, the addition every month or so of all sorts of eminently useful devices left Grumman in the position, so to speak, of having to wedge five pounds of candy into a four-pound bag. And since three-fourths of the LM's gross weight was accounted for by the pro-

---

* The original engine builder was Textron's Bell Aerosystems Company, but design problems caused NASA to bring in Rocketdyne as the backup contractor.

pellants that were required for its 140-mile round trip to
the moon, each pound of equipment that was added became
a four-pound problem. That is, for every additional pound
of equipment there had to be three more pounds of fuel
to move the heavier craft.

"We reached the point," Gavin said, "where we had to
say, 'Look, we've got to stop the design as it now stands
and squeeze some more weight out of it.' That is a very
embarrassing thing to have to do in terms of delivery
dates and costs, but we had no choice. We could see that
if nature took its course we'd have had a vehicle that would
simply have been too heavy."

Grumman had to put the LM through two weight-
reducing drives—SWIP, for Super Weight Improvement
Program. Eliminating parts, redesigning others to be
lighter and shaving the outer skin down to about the
thickness of heavy-duty aluminum foil—all this cost
some $30,000 per pound saved. And because the LM
would operate only in the moon's gravitational field, other
weight savings were possible. For example, the landing
gear could be lighter because the LM would fall slower
and softer in the low lunar gravity than if it were coming
down on earth. Likewise, the ladder down which the
astronauts would climb was built so lightly that it would
buckle under their weight if they used it at Cape Kennedy;
but on the moon, because of the low gravity, the men
would weigh a sixth of what they weighed on earth.

Just as Grumman was getting the LM's weight down, a
new problem developed unexpectedly. The launching-pad
fire that killed the three astronauts led to the addition of
100 pounds of fireproofing for the LM's cabin and electrical systems. If the fire had not virtually stopped the rest
of the Apollo program for nearly a year, the project would
have had to wait anyway for the LM to catch up. "It's
a terrible thing to have to say," a Grumman man told the
Times's Buckley, "but for us the fire was a blessing in
disguise."

To make the LM as fail-safe as possible, the engineers
tried every way to make the vehicle as simple as possible.
This presumably reduced the number of points at which
failure might occur. The rocket engines, for example, were
built to require no sparking device to fire. And, unlike
the liquid oxygen used in the Saturn 5, the oxidizer did

not have to be supercooled until it was used. The LM had no hydraulic or pneumatic systems, which might spring leaks, to cushion the jolt of landing; that was the reason for the crushable honeycomb in the struts.

With the exception of the ascent engine, all of the systems aboard the LM in which failure was thought possible were provided with duplicates—or backups. "We would like to have duplicated that [the ascent engine], too, God knows," a NASA official said. "But there just isn't room." Anyway, the engine was built as simply as possible, and many of the individual components had backups. In the event of a descent engine failure on the way down to the moon, the astronauts could jettison the stage in midflight and fire the ascent engine to return to the Command Module.

While there was only one computer on board for the critical guidance and navigation, its major circuits were built in triplicate and operated according to the principle of majority logic. If one of the duplicated circuits should start coming up with wrong answers, it would be overruled by the other two. That two should err was regarded simply and flatly as impossible. If it *should* happen, computers at the Houston control center would take over.

With these and other design considerations to worry over, it was no wonder that Joe Gavin concluded that the thing that surprised him most about the LM development "was the time it takes to do anything really well—it's much longer than you think." For this reason the first LM was a year late reaching the launching pad at Cape Kennedy for its maiden flight test.

Apollo 5

On Monday, January 22, 1968, a 16-ton unmanned LM surrounded by a protective shroud stood on top of a two-stage Saturn 1–B rocket. It was the same rocket that would have boosted Grissom, White and Chaffee into orbit. Now it was checked out and ready to send the Lunar Module into earth orbit on the mission designated Apollo 5.

Lift-off came at 5:48 P.M. after a four-hour delay caused by a rash of mechanical problems on the launching pad. Soon after the rocket's second stage, with the LM attached, reached orbit, the aluminum shroud covering the LM during the launching phase sprang open and was jettisoned.

"WE REACH THE MOON"

The man-to-the-moon epic introduced the world to gadgets and feats known previously only in Buck Rogers science fiction. In this section, stages of the story are documented in pictures chosen from the NASA archives.

President John F. Kennedy set the course in 1961 by committing the nation to a moon landing "before this decade is out." Within less than a year, astronaut John H. Glenn became the first American to orbit the earth.

3-2-1-0-Blast off! A new space-age suspense drama dawns with

the Mercury launchings at Cape Canaveral (renamed Kennedy).

At first the spacemen looked like visitors from another planet.

Early experiments with chimps reduced later risks for spacemen.

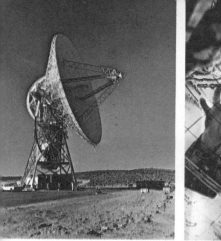

Tracking, plotting, communicating—all are vital for safe return.

Every move was followed at the Mission Control Center, Houston.

A new breed of folk-hero —John Glenn after his pioneer flight.

The Mercury flights, with stress on photography, were designed

to show that man could survive in a weightless state.

Pictures of cloud formations provided important weather data.

Walter Schirra with thermometer—a symbol of the health factor.

Buck Rogers? No, Edward White taking America's first space

"walk" during Gemini 4, with the earth as a dramatic backdrop.

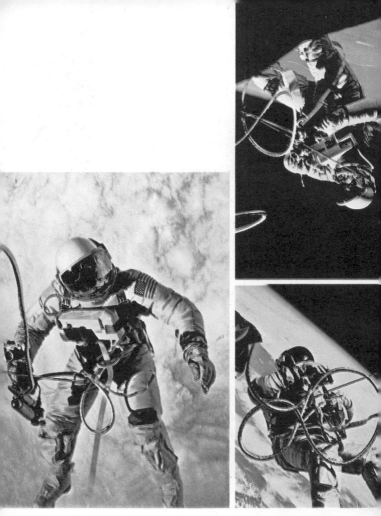

Space "walks"—with the astronaut at the end of an umbilical cord—tested man's ability to perform in hostile environment.

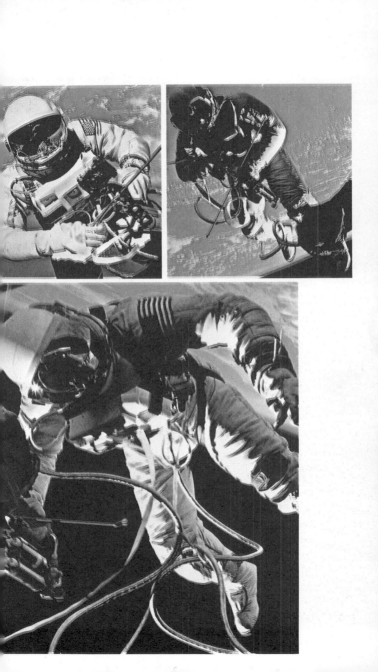

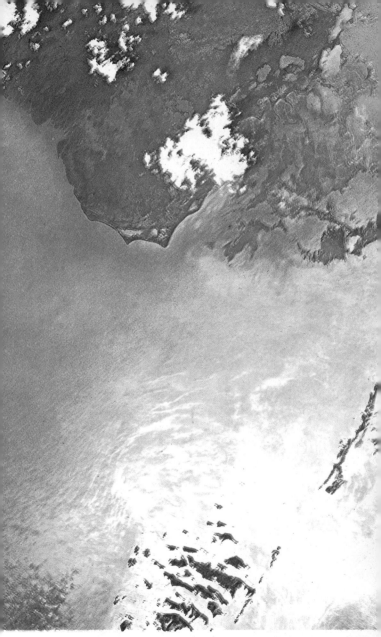

Some of the shots brought back by the high-flying Gemini pilots

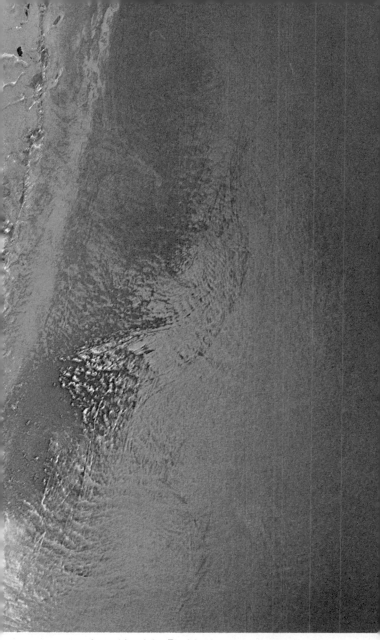

were rare—such as this of the Florida Keys and the Everglades.

Left: Titan 2 rocket; Control room, at Manned Spacecraft Center.
Below: After splashdown there was always the risk of drowning.

To get the astronauts out of the tossing waves as soon as pos-

sible, they were hoisted to hovering recovery helicopters.

Safe aboard a recovery ship, Cooper and Conrad receive the traditional red-carpet welcome.

A rubber collar keeps a bobbing capsule up.

Thomas Stafford during his pre-flight tests.

The first space rendezvous — achieved by Geminis 6 and 7.

Astronaut watches prelaunch activities (reflected in helmet).

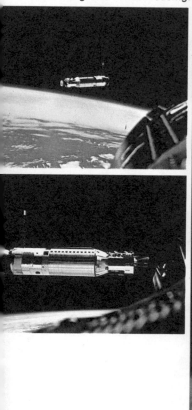
Agena, target for Gemini tests of docking and rendezvousing.

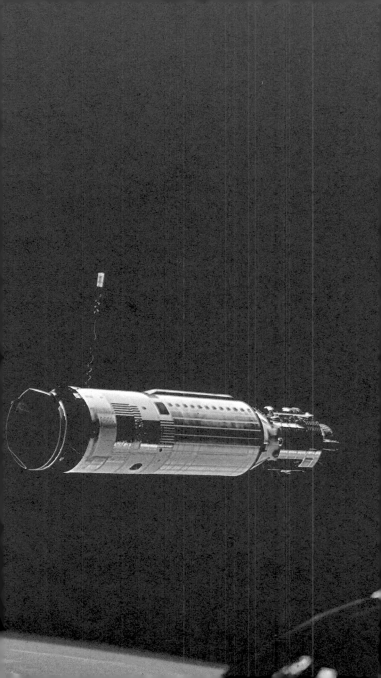

Conrad and Cernan at work in cramped quarters of Gemini capsule.

"An angry alligator" is what astronaut Thomas Stafford called this Agena target vehicle when the protective shroud failed to break free.

Time exposure of downward launch tower swing at Titan blastoff.

Joining vehicles — docking — was essential for return from moon.

The frogman's job was to speed water recovery.

John Young smiles up at a recovery helicopter.

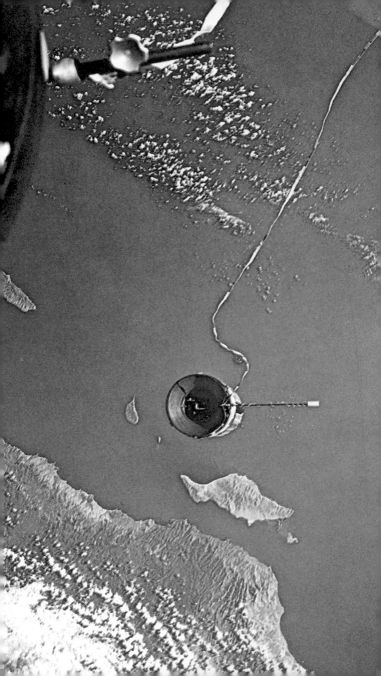

A tricky maneuver outside Gemini capsule.

An Agena docking target attached to Gemini.

Practice in weightless
state developed manual
skills for moon landing.

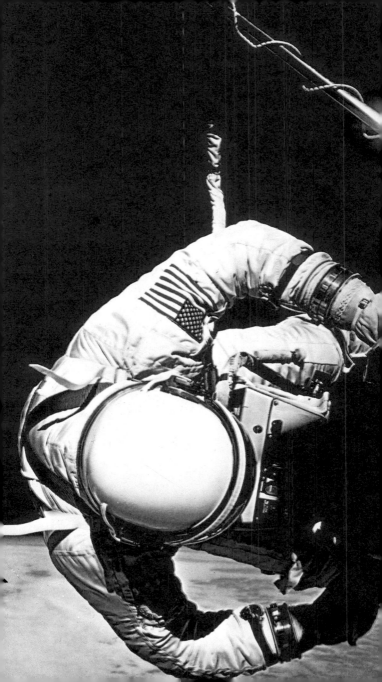

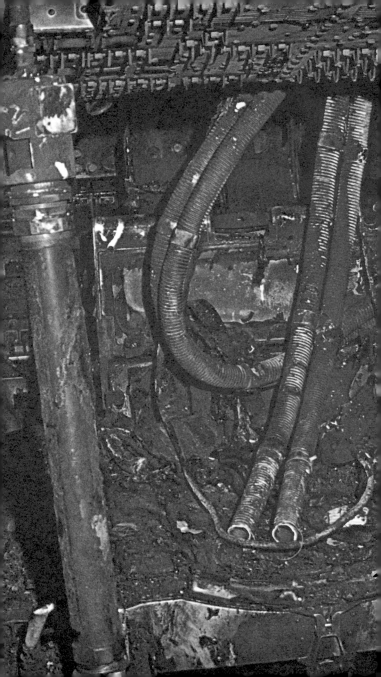

Suddenly, on Jan. 27, 1967, a catastrophic setback:
fire swept through Apollo 1 on the launching pad.

Three astronauts died in the tragic disaster:
Virgil I. Grissom, Edward White, Roger B. Chaffee.

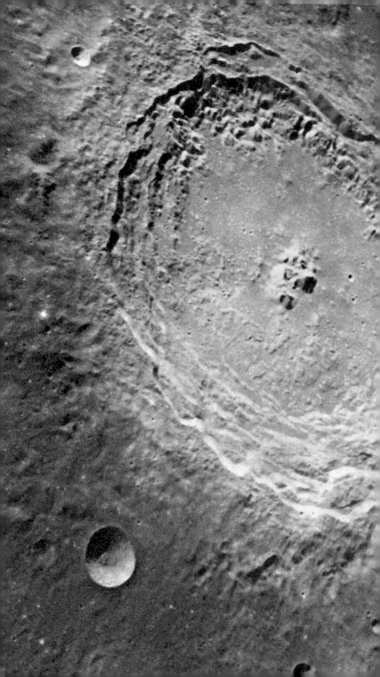

Pictures of Langrenus and other craters
helped determine the best landing sites.

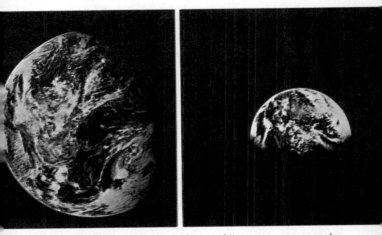

Looking back as Apollo 8 sped toward the moon—progressive
views of the planet earth as only space-age voyagers can see it.

As the men of Apollo 8 began their historic lunar orbit, it

was as if earth were peeking over a wall to cheer them on.

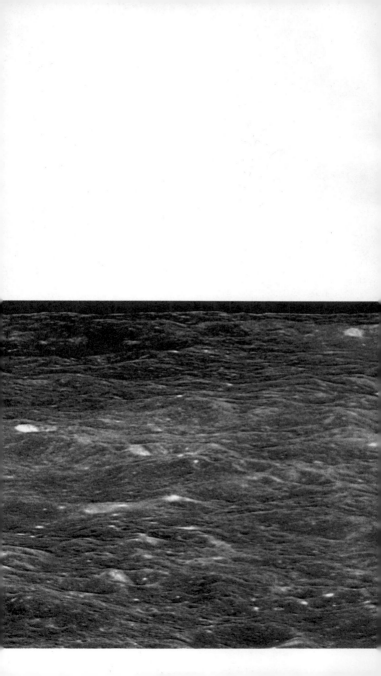

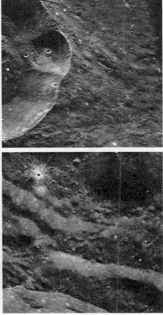

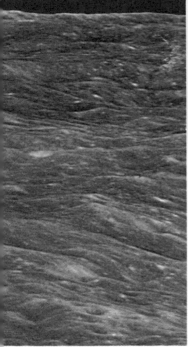

Lovell: "Looks like plaster of Paris or grayish beach sand."

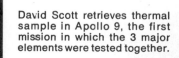

David Scott retrieves thermal sample in Apollo 9, the first mission in which the 3 major elements were tested together.

Maneuvering outside the LM.

LM's job: the delicate landing.

LM "spider" ready to descend.

Rainbow-like reflection from within the command ship.

Close-up of ridges and craters chart way for lunar landing.

We reach the moon! Astro Buzz Aldrin walks on lunar surfac

Picture by Neil Armstrong, first man to walk on moon.

Aldrin plants a solar wind collector for new information on the sun.

Away from Eagle, Aldrin sets up Passive Experiments Packag

which began at once to send back data on moonquakes.

And so, man has finally walked on the moon.
What new solar mysteries will his feat unfold?

Then the LM, acting on commands stored in its onboard computer, separated from the rocket, pulled away and began the first of two test firings of the descent engine.

But something was amiss. The rocket should have fired 38 seconds at bursts of power varying from 10 percent to full strength. Instead, the rocket cut off abruptly after a weak four-second firing.

While flight engineers puzzled over what went wrong, Eugene F. Kranz, the 34-year-old flight director, reshuffled the mission plan and radioed a new set of commands to the LM's guidance system. On the second try the descent rocket fired at full strength. But because of the earlier failure it was impossible to run a full simulation of the rocket's braking power during a moon landing. Later in the eight-hour mission, the ascent engine was ignited for a perfect one-minute test.

An analysis of radioed data from the LM indicated to ground controllers that the descent engine's initial shutdown was caused not by a failure but by oversophisticated equipment. Sensing a slow buildup of power, the LM's guidance computer suspected a rocket failure and switched off the engine. If men had been on board, flight controllers explained later, they could have immediately detected that it was a false alarm and reignited the rocket. "We have a very intelligent computer in our system," George E. Mueller, NASA's manned spaceflight director, said the next day. "But unfortunately it has a very rigid interpretation of the mission rules, and in this case the mission rule says you've got to get up to thrust in this period of time or else there is something wrong."

The Apollo 5 test ended after eight hours, and the LM remained in earth orbit, eventually to drop into the atmosphere and burn up. Apollo officials concluded that the mission had achieved 32 out of 33 objectives and was, therefore, a success. The Lunar Module, for all its delays and funny appearance, had passed its unmanned test and would next be called upon to carry men.

But the LM had not had the last of its weight and technical problems. Throughout 1968, engineers had to work on trimming excess pounds from its squat body and overcome a host of small wiring and structural troubles (see Chapter XV). It would be more than a year before men could try out the nation's moon-landing craft.

Chapter XII

The Command Ship

While the Saturn 5 and the Lunar Module were undergoing their debut flights, thousands of other Apollo engineers were busily putting the finishing touches on the Command Module, the ship in which the astronauts would leave earth and return. It was a critical and dramatic period, because the Command Module had been rebuilt to meet the rigid standards imposed after the launching-pad fire that killed the three astronauts in January, 1967. Were all the problems eliminated?

Everyone looked anxiously to the Apollo 7 flight, the first test of the modified Command Module and the first manned mission since the Gemini series had ended nearly two years before. For the rebuilding process had begun against the background of paralyzing self-doubts after the fire. But action soon replaced the self-doubts as tough schedules were instituted to produce fireproofing modifications that eventually would cost $400,000. In the spring of 1968, as the final product emerged, Dale Myers, North American Rockwell's Apollo program manager, said, "It's been a rough year for many people, but the program is really moving again."

The chief rebuilder was George M. Low, a Vienna-born, American-educated engineer who had held many technical and administrative posts in NASA since its beginning. In April, 1967, after the fire, as part of a high-level reorganization in the project, the 40-year-old Low was handed the toughest assignment of his career. He became the Apollo spacecraft manager, the man with overall responsibility for building a safer command ship for the moon voyage.

"Manned spaceflight is very unforgiving," Low said, as the project moved along. "We once made a serious mistake, a mistake of not maintaining absolute control over all flammable materials, and it took the lives of three of the finest men I have known. Since then, we have made every conceivable effort to avoid similar mistakes in the future. We have reexamined every drawing, every circuit and every component of the Apollo spacecraft. We have made thousands of changes in design, in manufacturing techniques and in tests. And we have literally rebuilt every Apollo craft."

The responsibility and burden were immense. Low directed a personal staff of 400. He could call on any of the 3,500 other men and women working at the Manned Spacecraft Center, and he had to oversee the work of at least 100,000 more persons employed by the contractors building the spacecraft and its components. All this meant that troubleshooter Low had to work a 90-hour week and spend much of his time airborne, skipping between space centers and the contractors.

After the Lunar Module flight in January, 1968, I spent a few days at Cape Kennedy to see what changes had been made in the year since the fire. New fire-fighting gear had been installed at the launching pads. Fire drills were frequent. Evacuation routes were marked by yellow arrows, and there was a long steel cable that would permit anyone in serious danger to slide to safety strapped to an inverted T-bar. "We can't eliminate all the risks," John Atkins, the Cape's new safety director told me. "Risk is the name of the game."

At the Houston Manned Spacecraft Center, engineers were testing all materials in the spacecraft for flammability under the most extreme conditions. Fires in the Apollo mock-up were touched off electrically, then watched by closed-circuit television and also monitored by instruments. Unless the fires extinguished themselves or did not spread, the materials were declared unacceptable. Such fires occurred at only five spots in the Command Module, and those items were already being redesigned.

I then went to visit the North American Rockwell factory at Downey, a sprawling community in the Los Angeles area, where the efforts of the company's 25,000-man space division were focused on getting "Wally's ship" ready for the first manned Apollo flight. Workers gave the vehicle that name because Walter M. Schirra, Jr., commander of the Apollo 7 crew, was personally supervising nearly every step of the craft's construction. Schirra and his crewmates, Walter Cunningham and Donn F. Eisele, had been the backup pilots for Grissom's ill-fated crew.

The man who was brought in as taskmaster for the Apollo 7 Command Module was John P. Healey, a 45-year-old engineer who had spent 27 years on aviation and missile jobs for Martin Marietta. On the blackboard in his office, among the chalked scribblings of schedules, one strange word stood out—"omphaloskepsis." It was the Buddhist term, Healey explained to me, for meditation while observing one's navel. It was a reminder, he said, that more hard work, and less introspection, was needed to get Apollo back on the track.

"The first thing I did was look at the people," Healey told me as we toured the plant. "Were they qualified? After 27 years in this business you can tell. I decided that the only thing they needed was what I call a more finite sense of direction. So I deliberately gave them some tough schedules."

One worker told Healey he could not finish a plumbing job in less than 24 hours. "Do it in two hours," Healey said, and the enraged man did it in two hours. Another job—the bonding of 54 brackets to the spacecraft floor—had taken pre-fire crews 16 days; this time around, the crews told Healey they could do it in eight days. Healey gave them 5½, and they did it in four.

Besides speeding up the work, Healey sought to impose greater control over the quality of the work. NASA investigators had charged North American Rockwell with some cases of poor management and inferior workmanship. Consequently, top supervisory engineers were assigned to all shifts, not just the daytime shift. A system of parallel work tasks was devised so that a single problem could not stop all progress. Healey also set up a production-control room, complete with closed-circuit television and wall-sized flow charts. With one look at the wall he could tell what key milestones had been passed, how long each had taken compared with how long it should have taken, and what should come next.

Such control was necessary if all the modifications in

the Command Module were to be made in time for the craft's scheduled shipment to Cape Kennedy in the spring. The changes included a new one-piece hatch that could be opened in seven seconds or less—compared with 90 seconds for the old three-piece hatch on the craft that had burned. Nearly all of the 1,412 nonmetallic items in the capsule were replaced by nonflammable materials or shielded by metal "firebreaks." The 20 miles of wiring were redesigned and insulated with Teflon wrapping, for it was suspected that frayed wires were the cause of the fire. Aluminum oxygen pipes were replaced in most cases with pipes of more fire-resistant stainless steel. Plumbing joints were "armored" so that damage during assembly and testing could not cause leaks.

Despite all the post-accident modifications,* a question still haunted the engineers, from Low to Healey and down the production line: Had anything been overlooked this time? "We've been searching down to the last wire for any other possible problems," Dale Myers said when I stopped by his office at North American Rockwell. Their search, he said, had been in vain. By late spring, the Command Module for the Apollo 7 earth-orbiting mission reached Cape Kennedy to prepare for the launching in the fall.

Simplicity and Redundancy

The modifications did not alter the original shape, size or weight of the Apollo Command Module. The original design grew out of discussions that had begun at the Langley Research Center in Virginia months before President Kennedy's Apollo decision in 1961.

Robert R. Gilruth, a leader in those discussions who became director of the Manned Spacecraft Center, recalled years later how some of the decisions had been made. Said Gilruth: "We had been working on key factors or guidelines for lunar flight. This was done in a series of bull sessions on how we would design a space-

---

* Most of the modifications were made not only on the Command Module but on the Lunar Module as well. The heavier plumbing added to the LM's weight problems during 1968.

ship for just circling the moon. Bob Piland was a key man
in this work along with Air Force Major Stan White,
Owen Maynard, Max Faget, Charles Donlan and a few
others of us. We got together evenings, weekends and
whenever we could to discuss such questions as crew size
and other fundamental design factors. We determined that
we would need three men on the moon trip to do all the
work required."

Gilruth conceded that there "were some who felt that
three was the wrong number psychologically, because of
the old saying, 'two is company and three is a crowd.' "
Nonetheless, three it was. The group also determined that
the vehicle would have to have onboard navigation capabilities, controlled reentry, an atmosphere so that the
crew could take off their spacesuits during the journey,
and the ability to fly a maximum of 14 days. These were
the "guidelines for lunar flight" which Gilruth presented
to NASA. "They have not required any change even to
this day," Gilruth said in 1969.

The basic design called for building the spacecraft in
three parts—the Command Module, the Service Module
and the Lunar Module. "We had designed our spaceship,"
Gilruth explained, "to have the Command Module on
top so that the astronauts could escape by means of the
escape tower if an abort were necessary. The Service
Module was underneath it with its big rocket for space
propulsion. The bottom element of the spacecraft was a
mission module to which the crew could transfer for
earth-orbital experiments. When the full landing mission
came along, we were able to substitute the Lunar Module
for the other module."

The Command Module's cone shape was derived from
the concepts suggested by Maxime A. Faget for the
Mercury and Gemini spacecraft. Such a shape gave the
craft a blunt end which, if pointed in the direction of
flight, would gain the maximum braking benefit from
atmospheric drag during reentry. Faget, who had been a
Langley man since 1946 and possessed what associates
called a "rare gift as a designer," also determined the
original internal arrangement of the capsule, working
closely with Caldwell C. Johnson, another Langley engineer and a skilled draftsman.

But these were only the bare outlines of the Command

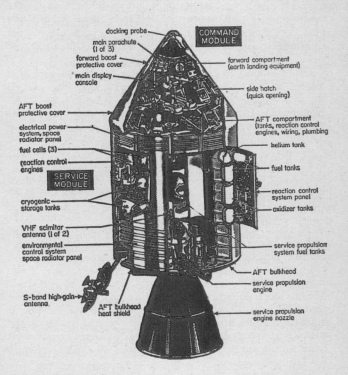

docking probe

COMMAND MODULE

main parachute (1 of 3)

forward boost protective cover

main display console

forward compartment (earth landing equipment)

side hatch (quick opening)

AFT boost protective cover

electrical power system, space radiator panel

fuel cells (3)

reaction control engines

AFT compartment (tanks, reaction control engines, wiring, plumbing

helium tank

fuel tanks

SERVICE MODULE

reaction control system panel

cryogenic storage tanks

oxidizer tanks

VHF scimitar antenna (1 of 2)

environmental control system space radiator panel

service propulsion system fuel tanks

AFT bulkhead

service propulsion engine

S-band high-gain antenna

AFT bulkhead heat shield

service propulsion engine nozzle

The Command Module, with room for three astronauts, sits atop the complex Service Module, which contains the Propulsion system.

Module. To be a living, working vehicle, it had to have wiring, electronics, pipes and valves, oxygen tanks and rockets. And to be reliable, it had to be as simple as such a complex vehicle could possibly be.

The early decisions made by Gilruth and his designers, therefore, provided for "simplicity and redundancy." All the rocket engines in the spacecraft, from the tiny maneuvering thrusters to the powerful Service Module rocket, used hypergolic propellants—that is, they were self-igniting on contact and consequently needed no spark plugs or ignition systems. Engine-thrust chambers and nozzles were made of ablative materials that melt and burn away, thereby avoiding the need for a complex cooling system. Engine valves were duplicated so that if one failed, the other could keep the engine running. Three independent fuel cells were provided, although one cell alone could generate enough electric power to return the ship safely from the moon. Nearly every point-to-point wire connection was duplicated, along a different path, though only one was required to do the job.

In short, George Low said, the design philosophy was to "build things simple, and then build two of them so that if one fails, the crew can still return to earth."

Profile of the Moonship

The Command Module (CM) and the Service Module (SM) were, in many ways, a single spaceship. The CM could not survive long in space without the SM, and the SM had no other function than, as its name implied, to serve the CM. The complex nature of the two modules, along with the launch escape system, can be seen in the following descriptions.

Command Module.—Sometimes called merely the Apollo, this was a much more spacious vehicle than the two previous models of American manned capsules, the one-man Mercury and the two-man Gemini. Being inside the Mercury was like being inside a telephone booth. The Gemini was about as roomy as the front seat of a small European car. The Apollo CM had twice as much room as Gemini, or about the spaciousness of a station wagon.

The conical CM was 10 feet, 7 inches high and 12 feet, 10 inches in diameter at its widest point, which was its blunt end. Its interior was the control center, the office and kitchen, bedroom and bathroom for three men for an entire flight, except for the period when two men would be in the Lunar Module for the descent to the moon. The CM's walls were lined with a fantastically complicated instrument panel and consoles. Its cupboards (bays) contained a wide variety of equipment, such as the guidance and navigation electronics and the astronauts' food, water, clothing and waste-disposal facilities.

Crewmen would spend most of their time in the three contour couches. The astronaut in the left-hand couch was the spacecraft commander who normally operated the CM's controls. In the center couch was the CM pilot, whose principal duties were guidance and navigation. He would often work down in a well in front of his couch, where he could operate the sextant and telescope for navigation sightings. On the lunar mission, he would be the astronaut who would be left behind in the CM while the other two men went to the moon's surface in the Lunar Module (LM). The astronaut in the right-hand couch was the LM pilot, and his principal task was to monitor the spacecraft systems—the electricity and oxygen supplies, the fuel consumption and the communications.

Unlike the astronauts in Mercury and Gemini, the astronauts in Apollo could get up and move around. With the center couch folded, two men could stand at the same time. Two could sleep in sleeping bags hung hammock-like below the couches. The cabin was normally pressurized with pure oxygen at about five pounds per square inch, or about a third of sea-level pressure, and the temperature was maintained at about 75° Fahrenheit.

The CM had five windows: two side, two rendezvous and one hatch. The side windows, about 13 inches square, at the side of the left and the right couches, were intended for photography and observation. The rendezvous windows, about 8 by 13 inches, faced the left and right couches and permitted a view forward toward the apex of the cone-shaped vehicle; they provided visibility for steering the CM to its rendezvous and docking, or linkup, with the Lunar Module. The hatch window was directly over the center couch.

Toward the apex of the CM was the tunnel containing the equipment for linking the CM with the LM. The two moon-bound astronauts would crawl through the tunnel to get into the attached LM.

The Command Module's main structure had double walls. Some of the early drawing-board concepts, made when radiation was feared to be a greater hazard, involved such thick walls that Caldwell Johnson said the CM would have looked "like a storm cellar." As it turned out, the inner wall, or pressure shell, was made of airtight aluminum in a sandwich-like configuration. Between two aluminum sheets was a bonded aluminum honeycomb core. This made for a substantial structure at less weight than would have resulted with a solid wall. The outer wall provided the shield against the heat generated by the ship's return through earth's atmosphere. Made of steel and a coating of reinforced plastic, it varied in thickness from less than an inch near the apex to 2½ inches at the blunt end where the "heat loads" would be the greatest. Altogether the CM weighed about 13,000 pounds at lift-off— including the crew.

Service Module.—This was the indispensable but often overlooked attachment beneath the blunt end of the CM. No men would ever ride in it. It would never reach the moon and would be discarded shortly before reentry to earth. But the CM had to have the SM. Housed in the cylindrical Service Module were the main rocket for the spacecraft, the rocket fuel and most of the CM's "consumables"—the oxygen, water and electricity.

The SM was just as wide as the CM, but more than twice as long—24 feet, 2 inches. Without fuel, it weighed 11,500 pounds; fully fueled, it weighed 55,000 pounds.* Its relatively simple structure consisted of a center section, or tunnel, running the length of the cylinder, surrounded by six pie-shaped sections.

In the center was the SM's main rocket, which was built for North American Rockwell by the Aerojet-General Corporation. The engine burned a fuel mixture of hydrazine and unsymmetrical dimethylhydrazine, and an oxi-

---

* These weights, like the weights of all sections of the Saturn 5, the Lunar Module and the Command Module, would vary slightly from mission to mission—but only slightly.

dizer of nitrogen tetroxide—the same propellants that were used in the Lunar Module. It was a restartable engine— with a bell-shaped nozzle jutting out the rear end of the SM—that could generate a thrust of 20,500 pounds. The astronauts would depend on this engine to get into and out of lunar orbit. The SM rocket was actually twice as powerful as necessary, having been designed originally for lifting the CM off the moon if the direct-flight method had been chosen for a lunar landing.

The SM's various pie-shaped compartments contained the propellant tanks for the rocket, plus the three fuel cells that would generate electricity for the spacecraft by a chemical reaction between liquid hydrogen and liquid oxygen. The by-product would be water, which would be piped into the CM and used by the astronauts to drink and mix with their foods. Here also were most of the oxygen supply and some of the communications equipment.

On the outside of the SM's aluminum alloy walls were

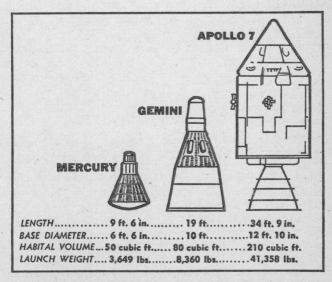

LENGTH.............. 9 ft. 6 in. .... 19 ft. .... 34 ft. 9 in.
BASE DIAMETER...... 6 ft. 6 in. .... 10 ft. .... 12 ft. 10 in.
HABITAL VOLUME...50 cubic ft...... 80 cubic ft....... 210 cubic ft.
LAUNCH WEIGHT.... 3,649 lbs........8,360 lbs.........41,358 lbs.

American spacecraft have grown bigger, roomier and heavier as man moved closer to the long lunar trip.

16 small rockets in clusters of four, 90 degrees apart. These made up the so-called reaction control system. They could be fired to make minor velocity changes, to maneuver during rendezvous wth the LM and to control the orientation of the spacecraft. The Command Module had 12 similar rockets that served as a backup system, but whose primary function was to orient the CM during its return through the atmosphere.

For the SM's fate was to be cast off as so much space junk about 15 minutes before the CM plunged into the earth's atmosphere for the splashdown. The unshielded SM would burn up coming into the atmosphere on its own. For those final minutes of flight without the SM, the astronauts in the CM would rely on batteries for electricity and a small reserve of breathing oxygen.

Launch Escape System.—For a short time after lift-off the Apollo spacecraft would carry a lattice-like tower rising above the Command Module's nose. This was the 33-foot-high launch escape tower. Its 147,000-pound-thrust rocket was ready, on command, to fire and pull the CM away from the rest of the spacecraft and the Saturn 5 in case of impending catastrophe. The rocket would shoot the CM out to sea and safety. In the event all went well with the launching, a smaller rocket on the tower would fire and jettison the unused safety mechanism.

Apollo 7

By October, 1968, all these complex pieces of equipment were ready for their first flight with men—Apollo 7. It was a crucial test. Had the designers and builders of Apollo corrected the problems laid bare by the fire? Had they committed some other tragic oversight? The setting for the start of the mission was the same Pad 34 on which the three men had died 21 months earlier. Any more serious trouble would all but wipe out American hopes of landing men on the moon before the end of 1969.

As October 11, the launching date, drew near, Cape Kennedy seemed to arouse itself to a tempo reminiscent of the earlier Gemini days. "It's good to be back in business again." These words were heard all week in the crowded bars and newsrooms, flight control rooms and offices. For the business at the Cape was firing rockets, and the biggest and most exciting business of all was launching a rocket with men aboard.

The rocket was Saturn 1-B, not as powerful as the big Saturn 5 but sufficient for the task of sending the Command Module and attached Service Module into earth orbit carrying three men. It was planned as an 11-day mission.

"She's riding like a dream," Walter Schirra, the commander, said shortly after lift-off at 11:03 A.M. "We're having a ball."

Schirra was a 45-year-old Navy captain and a veteran of Mercury and Gemini flights. He was born in Hackensack, New Jersey, the son of a World War I fighter pilot and a mother who took a few turns herself as a wingwalker in the barnstorming days. So it was perhaps natural that Schirra gained a reputation as one of the coolest and most unflappable of the astronauts. When his Gemini 6 spacecraft was poised for launching in 1965, the Titan 2 rocket fired, then unexpectedly shut down. Schirra had less than a second to decide whether to eject from the capsule or stay aboard and risk injury from a possible explosion. He quickly concluded—correctly, it turned out—that the safety device had not worked properly; by staying with the ship, he had saved it from being ruined in an ejection into the ocean. Asked moments later how he and his co-pilot, Thomas P. Stafford, felt sitting there atop a near catastrophe, Schirra answered: "We're just sitting here bleeding." But with his quickness also went a testiness and stubbornness that sometimes rubbed his closest associates the wrong way.

The man in the center couch was Donn F. Eisele, a 38-year-old Air Force major, methodical and somewhat introspective. (Eisele's first name was given to him by his mother who liked "Don" but added the second "n" so that it would not be confused with Donald.) He and the other crew member, Walter Cunningham, were rookies in space. Cunningham, a 36-year-old ex-Marine and ex-research scientist, was a bouncy, aggressive civilian who once said that "everything after my first lift-off is going to be pretty anticlimactic."

Over Hawaii, after orbiting earth nearly two times, the three astronauts cut loose the joined Command and Service modules from the second stage of the two-stage Saturn 1-B. This was the same S-IV-B rocket stage that on the bigger Saturn 5 would give men their final boost out of low earth orbit toward the moon. "If this were the lunar mission," Paul P. Haney, the mission commentator at Houston, explained, "that is approximately the point where we might ignite the Saturn IV-B to put us on a lunar trajectory."

The first two days of the mission were crammed with tests of the ship's electrical, navigational, propulsion and control systems. Using the spent rocket stage as a target, Schirra practiced rendezvous maneuvers to exercise both the Command and Service modules. Once, firing the big 20,500-pound-thrust Service Module rocket, Schirra exclaimed: "Yaba daba doo! That was a ride and a half."

A major objective of the mission was to perform eight firings of the big engine. All went perfectly, sometimes kicking the astronauts into an orbit as high as 277 miles, but always giving them a powerful kick. The fact that the rocket always fired on schedule, and for the exact amount of thrust, encouraged Apollo planners who were looking beyond to the time when the rocket *had* to work. If it didn't, men could be stranded in lunar orbit.

Apollo 7 was not without its niggling problems. A sensor gave a false warning on the craft's oxygen flow. Oversensitive circuit breakers temporarily shut down part of the craft's electrical system. The astronauts complained that some of the foods were too sweet, and the water often tasted too much of chlorine. For several days the crew, especially Wally Schirra, battled plain old-fashioned head colds. This perhaps accounted for some of the testiness in Schirra, who abruptly canceled the first of the mission's planned telecasts from orbit and several times sharply questioned orders from ground controllers.

But as Apollo 7 settled down to the routine of its long flight, the telecasts came to be highlights of each day. They ran from about seven to 11 minutes in length as the spacecraft passed over tracking stations at Corpus Christi, Texas, and Cape Kennedy. The "Wally, Walt and Donn Show," as it was nicknamed, was the first successful live telecast by American astronauts. The three men held up crudely lettered signs that read "Hello from the lovely Apollo Room, high above everything"—spoofing the time-worn radio opener used by band leaders broadcasting from hotels during the 1930's and 1940's. But for all the antics and corn, the world caught some brief inside glimpses of life aboard an orbiting spacecraft.

After 163 almost flawless orbits of the earth, logging more than 4 million miles, and taking hundreds of photographs of earth including the eye of a hurricane, on October 22 the Apollo 7 astronauts rode their Command Module down to a splashdown in the choppy waters south of Bermuda. Heavy swells tipped the spacecraft over, but the crew quickly inflated air bags to right the vehicle so that the apex, with its antenna, was pointed upward. Soon the astronauts were aboard the recovery ship, and Apollo officials back at Mission Control in Houston went into an orbit of their own.

101 Percent Success

Halfway through the Apollo 7 mission, a ground controller had radioed to Schirra: "Wally, you might be interested [to know] they're not even waiting for you to get back. They're simulating the next mission upstairs tonight."

And so they were, and some of the Manned Spacecraft Center officials made no secret of their wishes for the next flight: They wanted to send Apollo 8 around the moon.

At the traditional post-splashdown news conference in Houston, George Low could hardly contain himself. He had worked hard. So had thousands of others. They and their spacecraft were vindicated. "We accomplished 101 percent of the planned test objective," he said with pride. The extra 1 percent, Low explained, came from tests added to the flight plan while Apollo 7 was in orbit.

Apollo had passed out of the shadow of its darkest hour.

Chapter XIII

Communications and Control

When the Apollo astronauts entered space, they left behind them all the familiar landmarks that guide men on earth. There were no beaten paths, no signposts, no maps from the neighborhood service station. And because there was no gravity there was no way to tell what was up and what was down. There were only the stars. But where nothing else seemed to move, what would tell them how fast they were going? How could they find their way? How did the ground controllers keep track of the astronauts' movements and transmit instructions for their complicated maneuvers?

"If I had to single out the piece of equipment that, more than any other, has allowed us to go from earth-orbit Mercury flights to Apollo lunar trips in just over seven years," Christopher C. Kraft, director of flight operations at the Manned Spacecraft Center, said, "it would be the high-speed computer."

The computer, in all sizes and capabilities, was the coordinating brain in the vast system devised to guide, track and control man's voyages to the moon. Computers were used to check out the space vehicles before launching, and then to help carry out the launching itself. Scores of computers on the ground, strung around the world at tracking stations, processed the streams of radioed data passing to and from the spacecraft, information needed by flight directors to be sure of the spacecraft's condition and to know where the ship was, where it was going and where it had been. Because of the spacecraft's almost continuous radio link with earth, the flight director sitting before his control console in Houston was often called the "fourth man in the capsule."

During an Apollo flight, a miniaturized computer on board the Command Ship kept track of speed and position, calculated needed changes in the flight path, watched for malfunctions and displayed data on the cockpit panels for the astronauts' use. Because of such meticulous watchfulness, much of the driving in space was left to the onboard computer, the key unit in the spacecraft's complex guidance and navigation system. This is not to say the astronauts sat idly by, for they had scores of other tasks to perform.

In the Apollo 11 mission, the computer was crucial to the maneuvers involved in the actual landing, which, of course, had never before been attempted. Armstrong, Aldrin and Collins worked many long days in computerized simulators, practicing what they had to do to make the landing. Likewise, engineers spent nearly a year running make-believe landing flights over and over again to be sure the flight computer had all the right instructions for all the decisions and calculations it could be called upon to make.

G & N

No matter how powerful the Saturn 5, how sturdy the Command Module or how utilitarian the Lunar Module, it would not have been possible for men to fly to the moon without the Apollo's G & N—which was what the computerized guidance and navigation system was called. In fact, NASA considered the G & N so crucial that its development was the first major assignment given to a contractor after the Kennedy moon decision in 1961.

James Webb, the NASA administrator, turned to Charles Stark Draper, a professor at the Massachusetts Institute of Technology who had developed gunsights and navigation equipment for the Navy since World War II. As Draper recalled to me later, "Webb got me in there and said, 'Stark, this is going to be a hell of a job. Can it be done?' I said yes. We used to give lunar navigation problems as thesis work for our students, and I knew it could be done. It'll be ready before you need it, I told Webb—and, I might add, it is."

Draper then drove me in his 1961 Morgan, a decidedly non-space-age automobile he loved to tinker with on weekends, over to a converted hosiery mill on the Charles River. This was the center of activity for M.I.T.'s Instrumentation Laboratory, which had the prime contract to develop the guidance and navigation system under Draper's general supervision.

With a team of engineers, David G. Hoag, director of the Apollo guidance and navigation program at M.I.T., designed the system out of instruments already developed for the Polaris submarine missiles—a new custom-made computer. The AC Electronics Division of the General Motors Corporation was brought in to build the system under a $350-million contract. A small army of computer specialists working under Richard H. Battin, an associate director of the Instrumentation Laboratory and an authority on astronautical guidance theory, converted the Apollo mission goals and flight plans into a comprehensive computer program. This program was then recorded in the memory unit of a computer built for M.I.T. by the Raytheon Company.

The spacecraft computer, which was no bigger than a suitcase, was not like the ordinary general-purpose machine. It was designed for the special job of guiding an Apollo spacecraft, and would have been quite unable to process a single payroll check in a business office. More than 90 percent of the computer's 38,000-word memory was ineradicable. The memory unit consisted of tiny nickel-iron cores woven together like a rope by copper wires and encapsulated in plastic. Each core was a permanent instruction sheet for a specific mission; if, for example, Apollo flight directors changed an earth-orbital flight to a lunar-orbital flight, they would have to install a new computer.

The computer's core contained permanent data required for a particular flight: items like start coordinates, information about the position and gravitational fields of the sun and moon, and the equations necessary to make guidance calculations. To make such calculations, the computer had to have detailed, current information about the spacecraft's position, speed and altitude, or direction in which the ship was pointing. Such information came from three sources: an inertial measurement unit, or IMU, aboard the spacecraft; a ground-based computer at Mission Control that interpreted information from radar tracking stations, and sextant sightings of stars made by the astronauts.

The IMU, which looked something like a metal basketball, was the spacecraft's compass and speedometer. It provided the computer with a fixed reference direction by which to gauge the spacecraft's orientation—whether it was pointed toward the earth or toward the moon, tilted nose down or up, etc. It could also sense and relay to the computer any changes in speed and orientation that resulted from rocket firings.

Central to the functioning of the IMU was a metal platform suspended inside the metal basketball to allow free movement. Mounted on the platform were three gyroscopes, which were like spinning tops with a memory for a predetermined alignment. Any time the platform tried to tilt, the gyroscopes sensed the incipient movement and sent a signal to motors that kept the platform from moving. The platform thus remained stable, and the spacecraft in flight pivoted around it.

For the platform to be of use, the computer had to know how the platform's position was related to the star fields outside. It could find out from the astronauts. One crewman would use the spacecraft's sextant or telescope to sight on a given star. When he had it fixed in the center of the lens, he pushed a "mark" button that electronically transferred the sighting angles to the erasable part of the computer's memory—a sort of electronic scratch pad. He repeated the process with a second star. Since the computer already knew how the stable platform was aligned inside the spacecraft, it combined that knowledge with the star-sighting data and with the star coordinates stored in its permanent memory to produce a kind of space compass. The direction in which the spacecraft was pointed could always be measured with respect to the stable platform, which in turn was oriented to the stars.

Also mounted on the platform were three accelerometers, or space speedometers that measured changes in spacecraft speed. This information was converted to electrical pulses and transmitted to the computer, which stored that knowledge for use in calculations of future maneuvers.

Another source of information for the computer was the network of radar tracking stations on earth. The stations continually tracked the spacecraft to glean information about its range from the station, the rate at which this range was changing, and the spacecraft's direction of flight. After the information was processed by a ground-based computer, ground controllers could radio to the on-board computer the vehicle's speed and position at a certain time.

A third source of information consisted of star-and-landmark sightings made by the astronauts. Using the sextant, an astronaut could sight simultaneously on a given star and on a landmark or the horizon of the earth or moon. He would then push the "mark" button to enter the sighting into the computer memory. By making several such sightings in succession, the astronaut could tell the computer how fast the spacecraft was going and along what trajectory. This was essentially the same kind of information as the ground stations were providing, and was therefore gathered primarily as a double-check.

From this information, the computer could gain enough data to predict the spacecraft's future course and compare the prediction with the required course, and also to calculate any necessary engine-firings to alter the course. Such calculations were also made by computers at Mission Control and radioed to the spacecraft.

While the spacecraft computer was indispensable it was not autonomous. It could perform only certain functions, which were stored as programs in its permanent memory. Although it carried out its functions automatically once they began, the computer could not initiate a program without a direct order from the astronauts.*

Moreover, the computer was not foolproof for the actual touch-down of the Lunar Module for a moon landing. "The computer won't pick out the boulders for you," an official explained during a simulated practice landing. For this reason, before the Apollo 11 mission began, it was considered a strong possibility that the astronauts would switch from computer control to manual control at about 100 feet from the moon's surface.

If any of the G & N's major parts failed in flight, the astronauts would not necessarily have been in danger; they still could guide the ship manually. But because men cannot

---

* See Chapter XIV for a discussion of a typical Apollo day in space, which includes examples of how astronauts "talk" with the computer through a display keyboard. The primary guidance and navigation systems of the Command Module and the Lunar Module were essentially the same.

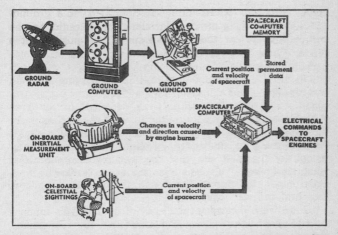

The Apollo guidance and control system. The on-board computer receives data from its own memory, radar stations, the inertial measurement unit and astronaut sightings to determine course and ship maneuvers.

carry out fast enough the millions of split-second computations required for accurate deep-space navigation, it would have been all but impossible to continue a lunar-orbit or lunar-landing mission. Thus, in the event of a G & N failure, the astronauts would have had to return to earth, using the manual back-up controls to fire the spacecraft rockets.

Keeping Track

Of all the vehicles and instruments developed for the moon mission, nothing quite compared in breadth and sophistication with the worldwide network of tracking antennas, communications relay stations and the flight control center that supported each Apollo flight.

The network's role was to enable the control center at Houston to communicate almost instantaneously with the spacecraft and its crew—except for a few unavoidable periods. These periods of LOS—loss of signal—came during the brief times that the spacecraft was between

network stations on its initial earth orbits, and for about
45 minutes of each lunar orbit, when the crew modules—
before and after separation—were circling behind the
moon. These times of LOS were some of the most sus-
penseful moments of any mission.

The tracking network consisted of 15 ground stations,
four instrumented ships and half a dozen or more instrumented aircraft.* The ships were deployed in the mid-
Atlantic and mid-Pacific to fill in communications gaps
left open where island sites did not exist. Besides all the
submarine cables, land telephone lines and microwave circuits, the network was tied together by communication
satellites hovering in orbit over the oceans, relaying messages from the spacecraft through the tracking station to
Houston and vice versa.

There were the voice messages, of course. This made it
possible for the astronauts to describe the view from space,
tell of their head colds and request instructions about
some instrument that was acting up. And the "voice link"
allowed ground controllers to read off new data for the
astronauts to feed into the computer and to indulge in
such chitchat as the baseball scores and the morning headlines. In addition, there were non-voice messages from the
spacecraft—coded data called telemetry. Telemetry involved the measuring of a quantity from sensors on board
the ship, transmitting the information as electric radio
pulses and then decoding it on the ground into meaningful
data about the flight conditions.

A NASA tracking station included one or more huge
steerable antennas. At the deep-space stations the antenna
dishes were as wide as 85 feet; at most of the other sites
the dishes were as narrow as 30 feet. The three deep-space
antenna stations (Canberra, Madrid and Goldstone) were
situated around the world so that, as the earth rotated,
always at least one would be facing the moon. Clustered

* The ground-antenna sites were: Antigua, Ascension Island, Bermuda,
the Canary Islands, Canberra (Honeysuckle Creek) and Carnarvon in
Australia, Grand Bahama Island, Goldstone in California, Guam, Guaymas
in Mexico, Hawaii, Madrid in Spain, Merritt Island in Florida, Tananarive
in the Malagasy Republic and Corpus Christi in Texas. Tananarive was
primarily a backup station for Apollo. Canberra, Goldstone and Madrid—
the deep-space stations for communications at moon distances—had two
antennas each, one primary and the other a backup. The four Navy
tracking ships were the *Vanguard*, the *Redstone*, the *Mercury* and the
Huntsville.

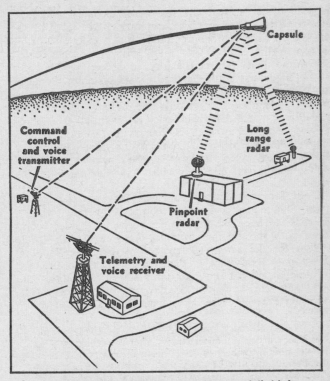

A typical tracking station in the worldwide network that helps
guide the space missions.

around the stations, in remote places so as to be removed
from distracting earth noises, would also be radar antennas
and an assortment of other whisker-like communications
fingers in the sky. More than 100 computers, built by the
Sperry Rand Corporation, were placed at the various sites
to process the data. The network, which during a flight
had 4,000 people on duty, was operated in large part for
NASA by the Bendix Corporation and represented an
investment of over $500 million.

At the network station near Tananarive off the coast of
Africa, the employees included, beside a cadre of American
engineers, some 120 Malagasy, which is what the people

of Madagascar are called. They were mechanics and technicians, diesel operators and secretaries. They considered the station their own, and certainly it was their one visible link with the Space Age. Many visitors to the station discovered that upturned empty spools on which cable had been wound could be used as picnic tables. Some of the visitors had read about the Apollo program, and more or less understood the station's role. Whatever their comprehension of the American space program, all the visitors quickly acquired an appreciation of the American hamburger. But the Malagasy, reported Lawrence Fellows, a *Times* correspondent, insisted on their own French word for the hamburger roll: They called it *pain de nasa*—NASA bread.

Even for sophisticated engineers long associated with space flight, it was hard to comprehend fully what a fantastic advance the Apollo network represented, compared with the original Mercury network. In an article written for the *Times* shortly before the moon landing, Christopher Kraft, one of the men chiefly responsible for the Apollo network, outlined the system's capabilities and why they were essential to the mission's success. The following excerpts were taken from Kraft's article:

In the Mercury program, we had relatively simple needs. Our chief concern right after launching was to determine whether or not the spacecraft had been rocketed into a satisfactory earth orbit and, if not, what actions we should take to accomplish a safe recovery of the astronaut. Once the Mercury capsule was in a proper orbit, the only flight maneuver that could change the flight path was the retro-fire maneuver that would finally slow the vehicle to bring it out of orbit and down to a safe landing near the selected recovery ship.

The problems of assessing whether the initial orbit was satisfactory, and of assuring safe recovery, seemed monumental at the time. But comparing them with the problems involved in carrying out the many combinations of maneuvers required on a lunar mission is like comparing simple addition to the most complex mathematical equation.

The communications that initially tied together the worldwide Mercury network consisted mainly of messages sent over low-speed teletype equipment. But that would not permit any direct voice contact between the

flight controller at what was then called Cape Canaveral
and the orbiting astronaut, except when the Mercury
passed within range of Canaveral. So it was necessary
that highly trained individuals take up residence at the
Mercury stations around the world at the time of each
flight to serve as flight controllers. By means of meters
and charts of data telemetered down from the spacecraft, these individuals (either astronauts or nonastro-
naut engineers) monitored the performance of both
spacecraft and astronaut, and relayed status information
via teletype back to the Canaveral control center.

Very early in the Mercury series we realized that the
15 minutes it took to absorb teletype data from distant
stations was too long and that it was vital to have
voice communications. We set to work to connect the
stations and the Canaveral center by high-frequency
radio. This type of gear was not very reliable, as you
will undestand if you recall the days when overseas
phone calls were made by radio instead of by undersea
cable or communications satellite. But HF radio sufficed
to provide the necessary ground-to-air control of the
Mercury vehicle, with its limited maneuvering capabilities.

The Mercury's successor, the two-man Gemini, was
designed to do a lot more maneuvering. It was vital,
therefore, that the ground control center have much
more immediate (what the engineers call "real time")
and more reliable data on the status of spacecraft and
crew. This meant connecting stations of the worldwide
network, first by cable and then by communications
satellite.

Finally, in designing the system that would support
Apollo earth-orbit and lunar missions, it was obvious
that the quality of the world network would have to be
advanced far beyond even Gemini's capabilities. That
meant using not only the latest high-capacity comsats
but also the high-capacity electronic computers needed
to digest and process the vast amounts of data that had
to be sent. Once such equipment was in operation, it was
no longer necessary to dispatch ground controllers to
each of the on-the-ground stations.

The progress made between the early Mercury days
and what we can do today is staggering. For instance,
during the seven minutes a Mercury was in range of the
station at Carnarvon, Australia, we were able to teletype
to Cape Canaveral data summarizing about 30 to 40
on-board functions. These might include heartbeat and

respiration of the astronaut, cabin temperature, oxygen supply, etc.

Today, with comsats and high-speed computers on board and on the ground, we can get almost instantaneous data at our Houston control center on as many as 500 gauges, dials or meters. The information from Carnarvon is received in Houston about two to six seconds after the Carnarvon station has received it from the Apollo capsule. The aero-medical doctor on duty can study the electrocardiogram from each of the three Apollo astronauts as though they were patients in his office. The doctor, or anyone else in the control center, can speak directly to the astronauts.

The computer used on the ground in the initial control center for Project Mercury was an adaptation of a computer that had originally been designed with no thought of space flight. It had been designed essentially for computations in scientific projects. It had a storage capacity of 32,000 words, which is 4,000 fewer words than the number compressed into each of the miniaturized computers carried aboard the Apollo Command and Lunar Modules.

The primary on-the-ground computer used on an Apollo flight has a capacity—an incredible contrast—of 5.5 million words. To the layman, this comparison may seem like a numbers game. But he would understand how really phenomenal the advance is if he considered how many thousands of man-hours were needed to fit together complex equations, involved logic and endless instructions into a workable set of computer programs for a lunar mission.

A Government-industry team has been working on these programs since 1962. There were times when the task appeared to be beyond our capability. But the challenge provided by the Presidential appeal to put an American team on the moon in this decade proved sufficient inspiration to the persons involved so that the programing task was finally accomplished. It is my opinion that the progress made by the nation's industry in computer technology—spurred primarily, I think, by the space program—is worth a large portion of the dollar outlay for the program.

Mission Control

The focal point for the Apollo ground network was the Mission Control Center at the Manned Spacecraft Center in Houston. (The Control Center first went into full operation during Gemini 4 in 1965. See Chapter VI.) It was a windowless, computerized communications facility. There were rows of consoles at which engineers and flight controllers watched coded data flicker across their screens. Unseen, in the basement below and in rooms to either side, were the banks of computers and telephone switches to handle the flood of information.

Mission Control received tracking and telemetry data from the network, processed it through its "real-time" computer complex and displayed the data at the consoles. For example, a data display would show the exact position of the spacecraft on a large map on a screen in front of the room. Any data indicating a problem would cause a red light to flash on the console of the responsible flight controller.

By the end of 1968, the sprawling Houston facility had grown to 90 buildings, with green lawns, a reflecting pool with ducks and wide parking lots. The men who worked there wore short-sleeve white shirts, thin dark ties, a small tie clasp and identifying cards clasped over breast pockets. It was almost like a uniform with them. Most of the men were tanned, crew-cut and young. Altogether, there were 13,000 secretaries and scientists, computer specialists and clerks as well as the astronauts and flight controllers.

"There is, you know, a kind of Walter Mitty quality about many of us here," Charles A. Bauer, a 49-year-old official in special events told Bernard Weinraub of the *Times*. "People like to feel they're on the team, part of the whole moon program. Even a secretary can say, 'The astronauts couldn't have done it without my help.' "

Added 49-year-old Andrew Sea, who was in the original task force that arrived at the center in 1962: "I'll tell you, this is an adventure, a once-in-a-lifetime thing. It's like being in Columbus's crew."

With the testing of the communications and tracking system, as well as the flight vehicles, during Apollo 7, the men at Houston's Mission Control were pushing the preparations for Apollo 8 in December. It would be the first flight taking men around the moon and close to the target of the decade.

Chapter XIV

Apollo 8: Around the Moon

By the time the space program was ready for the first long trip to the vicinity of the moon—with Apollo 8 in December of 1968, mankind was drained of hope, numbed with cynicism about events down here on earth. Much of the decade of the sixties appeared to be hard and bitter, especially so because it had started with such hope and vitality. Men deplored the present and feared for the future. Americans, in particular, were divided and demoralized.

Shock waves from the explosive year 1968 rocked the foundations of man's institutions. One quake followed another. Opposition to his Vietnam policies and general unrest had forced Lyndon Johnson to renounce another term as president, throwing the political camps into open warfare. Martin Luther King, who had come to be the conscience of the black man's quest for full equality, fell dead on a Memphis balcony from an assassin's bullet. His death triggered a riot of arson and looting in scores of American cities. It was a reaction of blind fury against the material side of a society that, through centuries of indifference and neglect, had called such a man into being and then killed him. The sad echoes of mourning had hardly faded when another man who had inspired hope among many, especially the young and the black, was cut down by another assassin's bullet. Robert F. Kennedy, brother of the slain president, was shot in Los Angeles as he sought the Democratic party's nomination for the country's highest office.

All the while the clangor of dissent rose in the streets and on campuses. At Columbia University, in New York City, students occupied the main buildings and in time forced the resignation of the university's president. In Chicago young people protesting the war in Vietnam disrupted the Democratic convention and, according to investigators, provoked a riot by the police. In France students and workers came very near to toppling the regime of Charles de Gaulle. In Czechoslovakia a trend toward liberalization became so threatening in the eyes of the Soviet Union that it invaded the country. Students went on a rampage in Egypt, Pakistan, Germany, Italy, Japan and Mexico. Even the pope was not immune as a rising number of Roman Catholics began questioning his authority over their private lives, especially whether or not they should use contraceptives. Discontent was epidemic in 1968.

The dark side of humanity seemed to be in the ascendancy when the three astronauts of Apollo 8 prepared to become the first men to circumnavigate the moon.

Decision to Go

Apollo 8's flight plan had been laid out months before. In August, 1968, General Phillips, the Apollo program director, announced at a news conference in Washington that the Lunar Module, because of check-out snags, would not be ready for testing on Apollo 8. To keep from losing time, the Apollo 8 mission was therefore changed from a manned earth-orbital test of the Lunar Module to a manned test of only the Command Module and Saturn 5 rocket. It would probably be conducted in earth orbit, Phillips said, but under questioning he conceded that, if Apollo 7 went well, Apollo 8 might be aimed for a flight around the moon.

Apollo 7 went well. And Chris Kraft and his associates at Houston, who had already worked out a detailed lunar-orbiting flight plan, applied all the pressure they could to get NASA headquarters to agree to such an assignment for Apollo 8. Thomas O. Paine, the new acting administrator of NASA, listened to all the arguments and on November 12 announced his decision.

"After a careful and thorough examination of all of the systems and risks involved," Paine declared at a news conference, "we have concluded that we are now ready to fly the most advanced mission for our Apollo 8 launch in December, the orbit around the moon."

The launch was to take place a few days before Christmas. Along the beaches and in the neighboring towns of Cape Kennedy thousands upon thousands of government officials, foreign dignitaries, aerospace executives and engineers, newsmen and tourists waited anxiously as the moment drew near. It was said to be the largest gathering at the Cape since John Glenn's launching.

Apollo 8 promised to be the most critical test thus far in the 7½ years since President Kennedy had initiated the Apollo project. American astronauts had flown 15 missions around the earth but had never ventured farther out than 850 miles. American robot spacecraft had crashed into the moon, landed on it and circled it; they had photographed it in detail, picked at its soil and analyzed its rocks. But now men were getting ready to go there themselves, to see the moon close up as no machine could— through their own eyes.

Although the Apollo 8 astronauts would not land on the moon, their flight would be a rehearsal of nearly all the other intricate maneuvers required for such a mission. It would also test the hardware—the rockets and spacecraft, computers and tracking antennas, launching facilities and recovery systems—whose design and construction had occupied roughly 350,000 people during the decade.

As the astronauts prepared to take off, a vast and far-flung network of men and ships and planes stood ready to chart the second-by-second events of their journey—to track them, watch over them, listen for their voices, pass on instructions for critical maneuvers, monitor the functions of crew and craft, keep tabs on their homeward course and, finally, pluck them safely out of the Pacific. The network embraced 14 ground stations, four heavily instrumented ships and six aircraft continually at the service of the lunar mission. The prime monitoring stations for deepest space penetration were in California's Mojave Desert at Goldstone, in Spain near Madrid and in Australia at Canberra.

Crew of Three

The commander of Apollo 8 was Frank Borman, the 40-year-old air-force colonel who was a veteran of the Gemini 7 endurance flight and who had served as the only astronaut on the fire-investigation board. Known for his rigid self-discipline, Borman neither smoke nor drank and his ability to concentrate had impressed all who worked with him. He was a devout Episcopalian who served as a lay reader in the church near the Houston Space Center. Eighth in his 1950 West Point class, he had returned there to teach thermodynamics and fluid mechanics and had earned a master's degree in aeronautical engineering at Caltech. Away from the job, Borman, the father of two boys, could be a relaxing and engaging man with a quick sense of humor, which he was to demonstrate later before Congress.

Borman's fellow crewmen were James A. Lovell, Jr., and William A. Anders. Borman and Lovell were no strangers to each other since they had shared Gemini 7 for 14 days. Lovell, a 40-year-old navy captain, became a close friend of Borman's during that flight and a convert to the Episcopal church. After his graduation from the U.S. Naval Academy, where he received a B.S. degree in 1952, Lovell served as a naval aviator, test pilot and flight instructor. In addition, he served on the President's Council on Physical Fitness. "I think I'd go wild," he once said, "if I had to punch a time clock and do the same thing every week."

Anders was the rookie. A 35-year-old air-force major with a master's degree in nuclear engineering, he had been born in Hong Kong to American parents. (Hong Kong newspapers played up the local angle during the flight, headlining HONG KONG MAN FLIES AROUND MOON.) Following in the footsteps of his navy father, Anders went to Annapolis, but then moved on to the air force as a pilot. Before joining the astronaut corps he had studied nuclear reactor shielding and radiation effects at the Los Alamos Atomic Laboratory in New Mexico.

Borman, Lovell, Anders—the names would soon resound around the world.

Waiting for Lift-off

On the night of December 20 the three were poised to embark on their journey, anticipating the greatest of Space Age adventures but mindful of the risks. In the weeks prior to launching a number of space critics had charged that Apollo 8 was not worth the gamble. To which Borman replied: "We've studied the mission, and we've studied the vehicle. We have faith in the guys who are helping us on the ground, and we have faith in the guys who built the machines. We wouldn't go if we didn't think the mission was worth the risks."

A successful mission would be an impressive space "first" for the United States. For a time it had appeared that the Russians would attempt to upstage Apollo 8. In September and again in November Soviet Zond spacecraft had looped the moon and returned to earth. They were unmanned but carried turtles, wineflies, mealworms and other biological specimens. Despite speculation, the Zonds were not followed immediately by manned flights around the moon.

Everyone at Cape Kennedy felt the mounting tension. "If we hadn't had other manned flights before," said Kurt H. Debus, director of its space center, "the excitement, the stress would be unendurable."

If the engineers, inured to countdown pressures, could be tense, so could a reporter. I must confess that I got no more than an hour or two of sleep the night before the launching. To make sure that nothing, no traffic jam or forgetful alarm clock, kept me away, I had gone out to the press site with blankets from my motel and was attempting to sleep on the floor of a house trailer *The New York Times* maintains there for covering space shots. The night was cool, slightly damp. The chill I felt was the kind of tingling goose-bump chill you used to have as a little boy when you and your family were leaving early in the morning on a long trip.

As the pale moon beckoned 220,000 miles away, technicians raced the countdown clock, fueling the rocket and overcoming pesky equipment problems, to meet the lift-off schedule at 7:51 A.M., December 21. Before dawn I gave up trying to sleep when I heard the voice of Jack King, the launch commentator, coming over the loudspeaker with the announcement that Borman, Lovell and Anders were on their way to Pad 39-A.

To the Moon

The thunderous launching was on time and flawless. The three-stage Saturn 5, rising out of billowing flames into the blue sky, slowly at first and then streaking, took 11½ minutes to boost the astronauts into a 118-mile-high orbit of the earth.

After completing nearly two orbits, in which they determined that the spacecraft was working perfectly, the astronauts struck out for the moon from a point over the Pacific Ocean south of Hawaii. At the time the area was in predawn darkness. People in Hawaii reported seeing the five-minute refiring of the Saturn third-stage rocket that kicked the spacecraft toward the moon.

"You're on your way," Chris Kraft, director of flight operations, radioed. "You're really on your way now."

"Roger, we look good here," Borman, the calm commander, told the control center.

About 30 minutes later the Saturn's third stage was separated from the spacecraft by the firing of explosive bolts. With it was jettisoned a dummy lunar landing craft that had been carried along for ballast. The astronauts were ordered later in the day to fire the Service Module's main rocket for two or three seconds—a "tweaking burn" as the engineers call it. Apart from correcting Apollo 8's aim, the firing was intended to give the astronauts confidence that the rocket engine, so essential in lunar-orbiting maneuvers, was functioning properly.

As the spaceship drew farther away from earth it gradually slowed down from its initial speed of about 24,200 miles an hour. It slowed because earth's gravity was still exerting a slight pull on the coasting craft. In a journey such as Apollo 8's a vehicle never actually leaves earth orbit until it goes into moon orbit. If the moon were not out there, Apollo 8 would have been in what amounts to a highly elongated orbit of the earth reaching a quarter of a million miles at its highest point.

This gave the astronauts an insurance policy. For if some problem forced them to cancel the lunar orbit, their course would take them around the leading edge of the moon, where lunar gravity would whip them behind the moon and back to earth without any major rocket firing.

On their second day in flight the Apollo 8 astronauts battled attacks of nausea but were still able to beam to earth live television that provided glimpses of life inside their moonship and a view of the bright ball of light that was the planet they had left behind. Darting past the windows, like tiny comets, were flashing specks of ice which streamed from waste water that froze after being vented from the spacecraft.

Lovell and Anders complained of occasional butterflies in the stomach. This was later attributed to a slight case of airsickness when they moved around too rapidly. Borman suffered an attack of what doctors thought was a virus ailment, commonly called intestinal flu. But before the day was out the three men seemed to be as healthy as ever.

At 3:30 P.M. on the third day, December 23, Apollo 8 crossed the great celestial divide. It passed out of the realm of space in which earth gravity dominates and into the moon's sphere of gravitational influence. The ship was traveling 2,216 miles an hour and was more than 214,000 miles from earth. It was the first time men had ever ventured into the gravity field of another body of the solar system.

Until this point the spacecraft had been traveling nose moonward. To get it into position for going into lunar orbit, the Apollo's maneuvering rockets were fired to turn the vehicle around so that its main rocket in the rear compartment was facing the moon. Thus the engine was so pointed that when it fired its exhaust would slow the spacecraft enough to drop it into moon orbit. The smaller maneuvering rockets were also fired to make a slight correction in the flight path. The aim was now nearly perfect.

Before crossing into the moon's sphere of influence, the astronauts had again focused their television camera on earth, showing half of it in bright reflected sunlight. It looked like a large misshapen basketball that kept bouncing around and sometimes off a viewer's television screen. From their vantage point 207,000 miles away, the astronauts said that they could make out the royal blue of earth's oceans, dark brown land and such configurations as Lower California and the mouth of the Mississippi River.

Speaking to his commander, Lovell said: "Frank, what I keep imagining is if I am some lonely traveler from another planet what I would think about the earth at this altitude. Whether I think it would be inhabited or not."

The astronauts could pick out even more detail than showed on the television pictures and Lovell attempted a description. "In the center," he began, "just lower to the

center is South America. All the way down to Cape Horn.
I can see Baja [Lower] California and the southwestern
part of the United States. There is a big cloud bank going
northeast, covers a lot of the Gulf of Mexico up to the
eastern part of the United States. It appears now that the
East Coast is cloudy."

In Lunar Orbit

A quiet excitement gripped the mission control room at
Houston on the day before Christmas. All the flight controllers, even those who would normally be off duty, gathered around the consoles to be on hand for the first anxious moment of lunar orbit.

The success and safety of the three astronauts hinged
on the performance of a single rocket engine. It had to
fire flawlessly to place Apollo in orbit and refire flawlessly
to send it back to earth. The bell-shaped engine, seven
feet wide and more than 13 feet long, sat in the aft end
of the 33-foot spacecraft's Service Module. It had been
fired once the first day out for a slight midcourse flight-
path correction and as a check that all was working
properly.

To reduce the chances of malfunction, the engine had
been designed with fewer than 100 moving parts and all
had emergency replacements—except the nozzle and the
combustion chamber itself. No ignition sparking device
was required since the fuel and oxidizer would ignite on
contact.

The engine, with its 20,500 pounds of thrust, had been
built by the Aerojet-General Corporation and tested thousands of times on the ground. An identical model had
fired successfully eight times on the Apollo 7 mission in
October. Since the moon's gravity is one-sixth as strong as
earth's, it is possible for such a relatively small rocket
engine to brake a spacecraft's fall toward the moon and
then put it out of orbit.

Early on the day before Christmas, Apollo 8 kept its
lunar rendezvous. As it raced toward the leading edge of
the moon, ground controllers checked the data on all of
the spacecraft's systems and decided to go ahead with the

crucial orbit maneuver. "Apollo 8," the controllers radioed, "you are riding the best bird we can find."

"Thanks a lot, troops," Anders replied. "We'll see you on the other side."

At 4:49 A.M., as the spacecraft curved behind the moon, the signals died out. Apollo 8 was out of range of the space agency's three deep-space tracking antennas. This was ten minutes before the vehicle's main rocket was supposed to fire, slowing Apollo 8 down and dropping it into lunar orbit. If the engine failed to fire, the astronauts would merely loop around the moon's back side, without going into orbit, and then whip back to earth.

But the engine ignited on schedule, at 4:59 A.M., when Apollo 8 was directly behind the moon, and fired slightly more than four minutes. It was pointed toward the moon at an angle so that its firing acted as an explosive brake, slowing the spacecraft from 5,758 miles an hour to 3,643 miles an hour. But it was 20 minutes more before the flight controllers could know if they had an orbiting vehicle. They waited in tense silence. Did the rocket fire? Was Apollo 8 in orbit?

When the spacecraft emerged from behind the moon a mission commentator in the control room exclaimed, "We got it! We've got it!"

Then came a trickle of data indicating that Apollo 8 was functioning well. Finally, a long minute later, there was a crackle of sound over the voice communication circuit. Lovell was talking. He was ever so matter of fact, the navigator first, leaving any poetry for later.

"Go ahead, Houston, Apollo 8," he said. "Burn complete. Our orbit is 169.1 by 60.5."

Amid the jubilation in the control room the message was acknowledged: "Apollo 8, this is Houston. Roger. 169.1 by 60.5. Good to hear your voice."

The numbers were the high and low altitudes of Apollo 8's moon orbit, given in nautical miles. This translates to about 194.5 statute miles at the highest point, which would be on the front side of the moon, and 69.6 miles at the low point, on the back side. Flight controllers and astronauts always deal in nautical miles.

Whatever the figures, the meaning was clear. Three men were orbiting the moon, journeying where no man had ever journeyed before, looking down on what Samuel

Butler had said were the "seas and lands Columbus and Magellan could never compass."

As the astronauts came around the eastern edge of the moon, traveling westward near the equator, the sun was shining high overhead. Most of the moon's back side and the eastern edge of the side facing earth were in sunlight. Much of the center part of the moon's face was partly illuminated by earthshine. The astronauts' first concern was not the moon but a radiator in the spacecraft's cooling system. All the water had evaporated and had to be replenished.

Then the astronauts turned their attention to what lay beneath them. One of the first major lunar features they spotted was Langrenus, one of many craters with peaks rising from the center of their floors. They next flew over the broad plain called the Sea of Fertility. When asked by ground controllers what "the old moon looks like," Lovell began describing the sights unfolding below:

> The moon is essentially gray, no color. Looks like plaster of paris or sort of grayish beach sand. We can see quite a bit of detail. The Sea of Fertility doesn't stand out as well here as it does back on earth. There's not as much contrast between that and the surrounding craters. The craters are all rounded off. There's quite a few of 'em. Some of them are newer. Many of them look like—especially the round ones—look like they were hit by meteorites or projectiles of some sort. Langrenus is quite a huge crater. It's got a central cone to it. The walls of the craters are terraced, about six or seven different terraces on the way down.

By then Apollo 8 had passed over much of the daytime areas of the moon and was reaching the Sea of Tranquillity, an even broader plain on the right side of the moon as seen from earth. With the sun lower, close to the horizon, the astronauts were able to make out more details and their perception of depth and height on the lunar surface was enhanced. This was the way they wanted it. For on the Sea of Tranquillity lay one of the five sites being considered for the manned lunar landing.

"It's about impossible to miss," Lovell assured the flight controllers. "Very easy to pick out."

The spacecraft then passed over the terminator—the

point where daylight changes to darkness. The view in that area, the astronauts reported, was quite sharp. But beyond, even with the earthshine, it became more and more difficult to make out any landmarks.

After moving around the back side and reappearing in their second orbit, the astronauts aimed their 4.5-pound television camera on the moon and began transmitting pictures at 7:29 A.M. It was the first of two TV broadcasts from the vicinity of the moon that were seen not only at the Houston control center but in millions of homes throughout the world—wherever television sets were available.

Throughout their orbit the astronauts gave many of the small unnamed craters on the back side the names of friends and associates. They even named three for themselves. The craters they dubbed Borman, Lovell and Anders lie just south of the equator near where the back of the moon ends and the front begins. Apollo officials said the crater names were in no way official, merely handy labels to identify some nameless features. The names were selected before the flight by the Apollo 8 crewmen and Harrison H. Schmitt, a geologist who is a scientist-astronaut. In addition to his doctorate from Harvard and Fulbright fellowship, Schmitt now had a lunar landmark bearing his name.

Some of the geographical christenings memorialized fellow astronauts who died in the line of duty, including the fire victims, Grissom, White and Chaffee. NASA officials who were honored included Webb, the former space agency administrator; Paine, acting administrator; Mueller, associate administrator; Phillips, director of the Apollo program; Gilruth, director of the Manned Spacecraft Center; Low, manager of the Apollo spacecraft program; Kraft, director of flight operations; von Braun, director of the Marshall Spaceflight Center in Huntsville, Alabama; Debus, director of the Kennedy Space Center in Florida, and Shea, the former Apollo spacecraft manager.

During the telecast Anders, who handled the camera, described the scene below: "The color of the moon looks like a very whitish gray, like dirty beach sand with lots of footprints in it. Some of these craters look like pickaxes striking concrete, creating a lot of fine haze dust."

At the end of the second complete orbit, when Apollo 8

was once more behind the moon, its main rocket refired. This dropped the spacecraft from an egg-shaped orbit to a circular orbit nearly 70 statute miles above the lunar surface. The rocket had again performed flawlessly.

The subsequent orbits were fairly quiet. The astronauts ate and took turns at some short naps. They continued to take color pictures and also practiced sighting landmarks used for navigation.

From the tracking data flight controllers noticed that the spacecraft tended to "jiggle" when it flew over the famous crater Copernicus. This was attributed to the fact, first deduced from the Lunar Orbiter flights, that under the moon's surface there are scattered lumps of material of greater density. A slightly greater gravitational tug is exerted at those spots. Some scientists, notably Harold C. Urey of the University of California, San Diego, have suggested that the moon is like a giant raisin cake with lumps of dense iron embedded in matter that is far less dense. Such a body could have been formed from a cloud of dust and larger objects, including chunks of iron, during the creation of the solar system.

If the moon were uniformly dense and perfectly spherical, the gravitational field surrounding it would be perfectly symmetrical. It was this field that held the Apollo spacecraft in orbit. The fact that the spacecraft's road was slightly bumpy, so to speak, revealed an uneven distribution of mass within the moon. The lumpiness of the moon is of concern because it makes less predictable an orbit for the descent of a spacecraft to the moon's surface —the goal of the Apollo program.

During Apollo 8's ten orbits of the moon it was observed, for example, that the spacecraft's flight path underwent some slight changes. After it was put into a circular orbit of 69.8 miles, Apollo 8 wound up sagging to nearly 68 miles at a low point and rising to nearly 73 miles at a high point. The wobbles of the lunar-orbit flight were apparently far too subtle to be noticed by the astronauts themselves.

While orbiting the moon the astronauts became the first men to witness a lunar sunrise and found it a strange and unexpected experience. According to Lovell, about two minutes before sunrise a fine white haze appeared over the horizon where the sun was about to appear. "It takes a fan shape," he said, "unlike the sunrise on earth, where the atmosphere affects it."

After being asked for further details Lovell said: "As the sun came above—or before the sun came above—the lip [the horizon], definite rays could be seen coming from the other side. It was a uniform haze, apparently from the center spot where the sun was going to rise. And this was something I couldn't explain."

His description, however, sounded strikingly like the corona, whose observation under these circumstances had been predicted. On earth the corona, consisting of rays radiating from the sun, is seen only at eclipse. This outrushing of gas extends beyond the earth as the so-called "solar wind," but it is dense enough to be seen only near the sun.

At about 9:30 P.M. on Christmas Eve the astronauts began their second and last television show from lunar orbit. It ran approximately 30 minutes and showed the bright moon, in a pitch black sky, outside the spacecraft window. Borman described the moon as a "vast, lonely and forbidding sight," adding that it was "not a very inviting place to live or work." Lovell saw the earth as a "grand oasis in the big vastness of space."

As the Christmas Eve telecast neared its end, Borman radioed, "Apollo 8 has a message for you." While the camera brought the startling vision of a desolate moon close up, Anders intoned the first words from the Book of Genesis:

> In the beginning God created the heaven and the earth.
> And the earth was without form, and void; and darkness was upon the face of the deep. . . .

Lovell then took up with the verse beginning:

> And God called the light Day, and the darkness he called Night.

Borman closed the reading with the verse that read:

> And God called the dry land Earth; and the gathering together of the waters called he Seas: and God saw that it was good.

After that Borman signed off, saying:

> Good-by, good night. Merry Christmas. God bless all of you, all of you on the good earth.

Return to Earth

After orbiting the moon ten times in about 20 hours, Apollo 8 headed back toward earth early Christmas morning. It fired its main rocket engine at 1:10 A.M. to kick out of lunar orbit and begin the 57-hour coasting voyage toward a splashdown in the Pacific Ocean.

Once again tension gripped everyone in the control room at Houston. For the rocket firing took place when the spacecraft was directly behind the moon and thus out of radio contact with earth. If the rocket failed to fire, the astronauts of Apollo 8 would remain in lunar orbit, doomed to die when their oxygen supply ran out. It was about 15 minutes before flight controllers got confirmation that all was well.

Through the static of 231,000 miles, as Apollo 8 swung around from behind the moon and started for earth, Lovell dispelled any doubts, saying, "Please be informed there is a Santa Claus." The rocket on which the astronauts' lives had depended, the 20,500-pound-thrust Service Module engine, had fired to perfection.

Leaving the moon they had orbited far behind, the Apollo 8 astronauts spent a quiet Christmas in space as they coasted back to earth at an ever-increasing speed. They recrossed the celestial divide between the moon's and the earth's spheres of gravitational influence. With the earth's gravity exerting a greater pull on the spacecraft, its maneuvering rockets were fired to adjust slightly the course it was following for a Pacific splashdown.

At 4:15 P.M. the astronauts again went on live television to show how they ate and exercised and how they guided and controlled their 33-foot-long ship. At the time Apollo 8 was traveling 2,929 miles an hour toward earth, 191,750 miles away.

Inside the cabin, Borman was seated in the commander's couch at the left side. Anders, the cameraman and systems engineer, was working by his couch at the right side. In the middle, Lovell's couch was folded back and he was lying suspended in the weightlessness. His head was out of sight down in the lower equipment bay. His feet could be seen strapped to a wall to keep him from floating around.

All three astronauts wore their soft strap-on hats equipped with microphones and earphones. They called them their "Snoopy hats," after the comic-strip beagle who often imagines he is a daring World War I fighter pilot. Instead of the bulky pressure suits worn for the lift-off, they wore the more comfortable flight coveralls made of lightweight Beta cloth, a fire-resistant material. On their feet were soft bootielike boots.

Lovell moved into the equipment area to demonstrate how the astronauts took their exercise. Lying on his back, he pulled two stretchable cords attached to a wall near his feet. "He's working with an exercise device that's designed to keep the muscles in shape," explained Borman, the narrator.

Next the camera focused on Borman and the display panel of the spacecraft's guidance and navigation computer system, which determined the direction, amount and duration of thrust required for all the spacecraft's rocket firings, including those that got the astronauts into and out of lunar orbit. "It's controlled by a DSKY [display keyboard], or similar to a typewriter keyboard," Borman said. "And the things that go in and out of that are absolutely miraculous." During the telecast the astronauts pressed various buttons on the keyboard to give the computer coded instructions for what they wanted to know or wanted done. "It's done a fantastic job for us," the Apollo 8 commander continued, "and Jim Lovell has done an excellent job operating it."

After the camera changed hands, Anders demonstrated the careful steps a hungry spacefarer had to go through to get a bite to eat. "The food that we use is all dehydrated," Borman said, "It comes prepackaged in vacuum-sealed bags."

Anders was preparing some hot cocoa. He cut a tip of the cocoa bag open with scissors and squirted five ounces of hot water into it with a pistollike injector. Each squeeze released ½ ounce. Then Anders passed the bag to another astronaut. And for good measure he handed over bags of sugar cookies, orange drink, corn chowder, chicken and gravy and "a little napkin to wipe your hands when you're done."

Borman described what Anders was doing to prepare some orange drinks: "You can see he's taking the scissors and cutting the plastic end off the little nozzle that he's going to insert the water gun into. And the water's going in. I hope that you'll have better Christmas dinners today than this."

The astronauts, however, had a more palatable meal than that shown on the telecast. It was a TV dinner of real turkey (not dehydrated), cranberry-apple sauce and a grape drink. In talking to ground controllers after the telecast, Lovell said that "we did the food people a grave injustice" by showing only the dehydrated food on television. He said the meal the astronauts ate was "delicious."

The next television shot showed Lovell at work down in the lower equipment bay that housed his navigation instruments. These included a sextant and scanning telescope with which the astronauts measured the angles from the spacecraft to selected target stars and lunar landmarks. Such readings are transmitted directly to the guidance and navigation computer to provide for automatic corrections in the flight path. During the day Lovell made a number of star sightings. "This is where we find out exactly where we are in space, what direction, and how fast we are traveling," he explained. "And our computer takes the information and tells how to maneuver to get home safely."

At one point in the day Anders asked a ground controller about his own Christmas and the latter said his son wanted to know "who's driving up there." "That's a good question," the astronaut replied. "I think Isaac Newton is doing most of the driving right now."

If not Newton, strictly speaking, it was the earth's gravity, which Newton explained centuries ago, that was drawing the men of Apollo 8 closer and closer to earth.

On December 26, their last full day in space, the astronauts' thoughts turned more and more to their native planet. During the afternoon telecast, their last of the mission, Anders mused: "I think I must have the feeling that the travelers in the old sailing ships used to have, going on a very long voyage away from home and now we're headed back. I have that feeling of being proud of the trip but still happy to be going back home." The three men spent most of the day carefully checking out their maneuvering rockets, charging batteries, stowing equipment and dumping waste

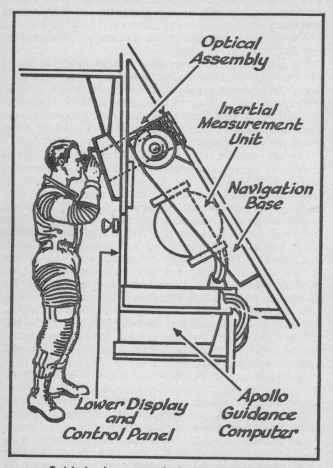

To take bearings on moon landmarks, astronaut uses
either sextant or telescope.

water in preparation for their return to earth—considered one of the riskiest phases of the mission. Flight controllers cautioned against talking of a successful mission before it was over.

Yet the return to earth on the morning of December 27 went off with the same precision that had characterized the entire trip. "It was all automatic," said Borman, shrugging off the congratulations of the pilot of the helicopter that lifted his crew out of the Pacific Ocean south of Hawaii. The spacecraft's guidance and navigation system had controlled the fiery dive into a 26-mile-wide atmospheric corridor at 24,530 miles an hour, faster than any other manned reentry into the earth's atmosphere.

By the time Apollo 8 had reached the upper layers of the atmosphere at about 400,000 feet, the cylindrical Service Module had already been separated from the cone-shaped Command Module in which the astronauts were riding. The bolts holding the two modules together were blown away by the firing of mortarlike explosives. The Service Module's small thrusters were also fired to pull it out of the Command Module's flight path. Then the astronauts were cut off from their in-flight storeroom and power plant. Inside the Command Module the computers and other systems were left to run on battery power and the astronauts, braced in their couches, were breathing oxygen from a reserve tank. What had started out as a 6.5-million-pound vehicle—the Saturn 5 three-stage rocket and the full spacecraft—was now only an 11,000-pound capsule.

Shortly after the Service Module separation, the onboard computer triggered the firing of six thruster rockets to turn the spacecraft's blunt end toward the line of flight. The men were thus riding down backward.

The outer surface of the Command Module was coated with a reinforced plastic called phenolic epoxy resin, from a half inch to two inches thick on the blunt end. This coating served as a shield against the intense heat that built up on the Command Module's outer surface when it slammed against the atmosphere. The material turned white hot from the friction, charred and melted away, thus dissipating nearly all the heat before it could penetrate the capsule. Meanwhile the capsule itself was rolled over slowly so that no one side would overheat during reentry. The amount of heat that had to be dissipated was staggering—about 5,000°

Fahrenheit a few seconds after the capsule hit the atmosphere. The intensity of the heat blacked out communications. But this had been expected and the blackout lasted almost exactly the three minutes that had been predicted. At two brief periods during reentry the spacecraft rose instead of continuing its descent. The purpose of these roller-coaster maneuvers was to help reduce the jarring impact on the astronauts and they were accomplished by short automatic firing of the thruster rockets.

While Apollo 8 was streaking at a slant across Siberia and northern China, Japan and then the mid-Pacific, the tracking ships deployed under the descent path were waiting to relay communications. Suddenly they began picking up messages from the spacecraft and then, to the relief of all, Lovell's voice could be heard through the crackle of static. "Looking good," he said.

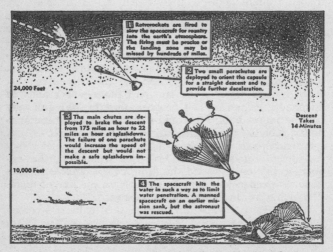

1 Retrorockets are fired to slow the spacecraft for reentry into the earth's atmosphere. The firing must be precise or the landing zone may be missed by hundreds of miles.

2 Two small parachutes are deployed to orient the capsule for a straight descent and to provide further deceleration.

24,000 Feet

3 The main chutes are deployed to brake the descent from 175 miles an hour to 22 miles an hour at splashdown. The failure of one parachute would increase the speed of the descent but would not make a safe splashdown impossible.

10,000 Feet

4 The spacecraft hits the water in such a way as to limit water penetration. A manned spacecraft on an earlier mission sank, but the astronaut was rescued.

Descent Takes 14 Minutes

Schematic drawing

How an Apollo spacecraft descends into the earth's atmosphere—key steps and the problems.

The first of a series of three parachute systems deployed automatically eight minutes after reentry began, when the spacecraft was 23,300 feet above earth. By then it had slowed to about 300 miles an hour. First to unfold were the two 14-foot-wide drogue chutes, which oriented the craft and continued slowing it. Less than a minute later Apollo 8 was dropping straight down. Then three small parachutes pulled out the three main parachutes, each 83 feet in diameter. When the nose broke away to deploy the parachutes it exposed a beacon. This flashing white light could be seen from the aircraft carrier *Yorktown* and the waiting helicopters as Apollo 8 hit the water at about 20 miles an hour—only four miles from the *Yorktown*.

With the first circumnavigators of the moon safe and in good health, a vast sigh of relief went through the Houston control center. The sigh was followed by pandemonium—backslapping, cheers and even a few tears. Small American flags were planted on the control consoles. A huge American flag was draped across the face of a huge electronically operated map of the earth that dominated the room.

"It's a veritable roar in here—the room is awash with cigar smoke," said Paul P. Haney, the space agency public affairs officer who had acted as the "Voice of Apollo" during the splashdown and other key maneuvers of the mission. He added that the "demonstration going on here . . . we haven't seen ever in our history."

"This flight is one of the great pioneering efforts of mankind," Paine said at a news conference. "We feel very humble that we were the ones that were given this opportunity to perform this historic feat."

Keynoting the new spirit of self-confidence that swept the center, Paine added: "This is not the end but the beginning. We are here this morning at the onset of a program of spaceflight that will extend through many generations. Man has started his drive out into the universe."

Hero's Welcome

After their return to Houston and eight days of exhaustive debriefing, the men of Apollo 8 were welcomed by a proud and awed people into the select ranks of national heroes.

They were awarded gold medals by President Johnson at the White House and paraded up New York's Broadway in a blizzard of ticker tape. They were hailed as "history's boldest explorers." Ten Soviet cosmonauts sent their own very special congratulations, noting: "Your successful orbiting of the moon on board the Apollo 8 spacecraft and a successful return to the earth are another milestone in scientific and technical progress."

Some could hardly believe what had happened. "I never thought this day would come," said a New York City housewife, "and now that it's here I still find it hard to believe." General Phillips, who had worked years to make the flight possible, found himself shaking his head and saying more to himself than to the reporters, "It's hard for the mind to comprehend. It's really hard for the mind to comprehend." Even the astronauts shared the sense of wonder. Lovell told congressmen: "I stepped out of the house a few days later and looked up at the moon and I could scarcely believe that I was there. A sense of pride, a feeling of satisfaction and achievement came over me. And I thought to myself, is there some American in this great country who, when he sees the moon, cannot feel the same as I do and say to himself, 'We were there'?"

Some still had doubts about the wisdom of the enterprise. Elaine Ford, chatting with some other Negro workers in a Detroit office, said, "There is something else they could have done with all that money, something they really need." But in New York upon seeing color pictures of the moon and of the earth as a small sphere, one skeptic of the Apollo program had a change of heart. "Now," he said, "I know why they want to fly to the moon."

In their public appearances Borman, Lovell and Anders emphasized that their mission had achieved all its objectives. By journeying more than 500,000 miles, farther out than man had ever before ventured, they had demonstrated that the Saturn 5 was as reliable as it was powerful, that Apollo's navigation and communications systems were equal to the lunar-flight challenge and that the Service Module engine could be counted on to fire with precision. Borman said that they had encountered no problems that "would be of any concern to future flight." The spacecraft, he said, "is a living, breathing, magnificent piece of machinery."

Lovell reiterated at a news conference in Washington that the Sea of Tranquillity was a "safe and adequate" site for landings. The astronauts were unable to investigate other possible landing sites because the sites lay in darkness during their orbits.

Not only did Apollo 8's tremendous success inspire new confidence in the moon-landing project but it brought renewed hope to Americans and mankind in general. It seemed to make a profound spiritual impact on people. In the aftermath *Time* magazine scrapped plans to feature "the Dissenter" as its "Man of the Year" and substituted the three astronauts. "For the American people," *Time* wrote, "the astronauts' triumph came as a particularly welcome gift after a year of disruption and despond."

The astronauts sensed this and shared the feeling. When he stood before a joint session of Congress, Frank Borman said that the most impressive moment in the flight had come when the three voyagers saw earth, a small shining sphere, rising over the lunar landscape. He told Congress that a prose poem by Archibald MacLeish "captures the feelings that we all had in lunar orbit." MacLeish's reflections were written expressly for *The New York Times* and appeared the day Apollo 8 was circling the moon. Apollo 8 moved him to write:

> Men's conception of themselves and of each other has always depended on their notion of the earth. When the earth was the World—all the world there was—and the stars were lights in Dante's heaven, and the ground beneath men's feet roofed Hell, they saw themselves as creatures at the center of the universe, the sole, particular concern of God—and from that high place they ruled and killed and conquered as they pleased.
>
> And when, centuries later, the earth was no longer the World but a small, wet, spinning planet in the solar system of a minor star off the edge of an inconsiderable galaxy in the immeasurable distances of space—when Dante's heaven had disappeared and there was no Hell (at least no Hell beneath the feet)—men began to see themselves, not as God-directed actors at the center of a noble drama, but as helpless victims of a senseless farce where all the rest were helpless victims also, and millions could be killed in world-wide wars or in blasted cities or in concentration camps without a thought or reason but the reason—if we call it one—of force.
>
> Now, in the last few days, the notion may have

changed again. For the first time in all of time men
have seen the earth: seen it not as continents or oceans
from the little distance of a hundred miles or two or
three, but seen it from the depths of space; seen it
whole and round and beautiful and small as even
Dante—that "first imagination of Christendom"—had
never dreamed of seeing it; as the twentieth century
philosophers of absurdity and despair were incapable
of guessing that it might be seen. And seeing it so, one
question came to the mind of those who looked at it.
"Is it inhabited?" they said to each other and laughed—
and then they did not laugh. What came to their minds
a hundred thousand miles and more into space—"half
way to the moon" they put it—what came to their
minds was the life on that little, lonely, floating planet:
that tiny raft in the enormous, empty night. "Is it inhabited?"

The medieval notion of the earth put man at the
center of everything. The nuclear notion of the earth
put him nowhere—beyond the range of reason even—
lost in absurdity and war. This latest notion may have
other consequences. Formed as it was in the minds of
heroic voyages who were also men, it may remake our
image of mankind. No longer that preposterous figure
at the center, no longer that degraded and degrading
victim off at the margins of reality and blind with
blood, man may at last become himself.

To see the earth as it truly is, small and blue and
beautiful in that eternal silence where it floats, is to
see ourselves as riders on the earth together, brothers
on that bright loveliness in the eternal cold—brothers
who know now they are truly brothers.

Chapter XV

Apollo 9: "The Connoisseur's Flight"

It was now 1969, early January in the year of Apollo's ultimate test. While the excitement of Apollo 8 was still simmering, a door in Cape Kennedy's mammoth Vehicle Assembly Building slid open and from inside another Saturn 5 slowly emerged, standing upright on a crawler-transporter. Seven hours later, the giant rocket completed its 3½-mile trip to Launching Pad 39-A. Atop Saturn was the spacecraft for Apollo 9.

A scheduled launching date of February 28 was established, and all preparations went smoothly; but as the day drew nigh project officials and engineers spoke in guarded terms about the flight. Apollo 9 would mark the first time all the major components for a lunar-landing mission—the Saturn 5, the Command Module and the Lunar Module—were tested together. Above all, it would be the first time astronauts got a chance to maneuver in flight the Lunar Module, the fragile, spindly-legged craft that on a real moon mission would shuttle men from the orbiting Command Module to the moon's surface and back. The Command Module had passed its test on Apollo 7, the Saturn 5 had passed its manned test on Apollo 8, and now the moment had arrived for the Lunar Module's manned debut. The planned ten-day mission, Apollo officials cautioned, would be the most complex and riskiest flight thus far undertaken.

Apollo 9's flight plan called for the astronauts to go into earth orbit where they would practice a series of maneuvers essential to the lunar landing. They would practice joining the Command Module with the Lunar Module, flying the combined craft and then separating to perform rendezvous exercises between the two. What introduced an added element of risk was the fact that the thin-walled Lunar Module, or LM, was incapable of returning to earth.

"I don't know if you've ever seen a tissue-paper space-craft before," James A. McDivitt, the Apollo 9 commander, said, "but this thing sure looks the part." Chosen to take the risk with McDivitt were David R. Scott, the no. 2 man, and Russell L. Schweickart, a space rookie.

The LM, designed only for operations on and around the airless moon, lacked any shielding against the tremendous heat generated when a spacecraft plunges through earth's atmosphere. So when astronauts entered the LM and cast off from the Command Module, their lives depended on being able to steer the LM back to the mother ship. "When you make that separation you've got to rendezvous," Christopher Kraft of the space center explained. "We've never been in a position in the program where we *had* to rendezvous, either us or the Russians. But now, when you separate that 'bear' you've got two guys off there in a vehicle that can only last just so long in the first place, and in the second place you can't reenter."

While the astronauts were contemplating Apollo 9, the Soviet Union sent two manned Soyuz spacecraft into earth orbit. One pilot was aboard Soyuz 4, launched on January 14. Three men were on Soyuz 5, launched the following day. The two ships joined each other in space and on January 16, two of the three men from Soyuz 5 "walked" outside to transfer from their craft to Soyuz 4. It was the first time men had transferred from one orbiting vehicle to another. But Apollo 9 had similar plans—though the transfer would be through a tunnel between the vehicles, not outside—and so Apollo project officials were not overly concerned with the Russian "first."

Indeed, Apollo officials had more than enough to think about with their own mission: the Lunar Module had been tested only once in unmanned flight (the Apollo 5 mission in January, 1968) and had a history of many pesky problems. Its failure to pass prelaunching inspections on schedule had delayed its first manned test from Apollo 8, as originally planned, to Apollo 9. In an article for *The New York Times* Sunday magazine, Tom Buckley, a staff writer, reviewed some of the problems after talking with Kraft.

As anyone who has ever broken a shoe lace and cut himself shaving while trying to get ready to catch a plane knows, efforts to save time end up the other way around, and so it was with the LM.

"LM 1, the only LM that has been launched so far, was highly successful," Kraft said, "but it wasn't fully equipped. It didn't have its life-support system, communications or radar. There wasn't any point in putting them on board since the mission was unmanned and there wouldn't have been anybody to use them."

Because of an error in the programming of its computer, LM 1, which was supposed to have continued in earth orbit long enough for its functioning systems to fail (either for lack of power or by exceeding their rated service span), re-entered the atmosphere prematurely and was destroyed. The LM 2 mission, which was also to have been unmanned, was canceled by NASA as a needless duplication of effort. The LM was sent to Houston where it was used in stress and vibration tests on the ground.

The Apollo 8 mission did not carry a Lunar Module, since it would have been destroyed for no purpose. Instead, containers of water with the same mass and center of gravity as the LM were placed in the LM compartment to test the lifting ability of the three stages of the Saturn.

Although such relatively crude articles of commerce as automobiles and television sets are made of parts that can, for all practical purposes, be regarded as identical with other parts, Kraft went on to say, this is not true of the components used in the space program. The very fact that the metal for a certain bolt has been selected with greater care than is a dam for a Kentucky Derby winner, and has then been machined to a tolerance of a five-thousandth of an inch, and that perhaps 50 or 500 bolts have been inspected and X-rayed and rejected along the way, tends, paradoxically, to create a mechanism that is highly individual and may well be full of quirks.

Not only are no two items entirely identical, Kraft said, but they also are likely to affect one another in an almost metaphysical way once they are installed in a spacecraft. "There's a lot we still don't know about these interactions," he said. "Problems have to be solved through 'cut and try' rather than by theory." Adjoining wiring systems "talk" to each other and have to be re-routed. The rendezvous radar on the LM 3 [the vehicle for Apollo 9] . . . performed perfectly in factory tests, but once installed, set up electrical interference with other systems. A cabin-lighting control panel turned out, Kraft said, "to be a pretty good little radio receiver." It had to be removed and rebuilt while the schedule continued to slip. "It's like ripping the nervous system out of your body and then putting it back," he said, grimacing with remembered pain.

Microscopic examination disclosed tiny cracks in the soldered joints of the tubing that carried the fuel and oxidizer to the rocket engines, and X-rays showed the presence of minuscule deposits within. It hardly seems conceivable that either condition could affect the operation of engines whose service life will be measured in minutes, but an entirely new thin-metal welding technique was developed to replace the solder.

The slipperiest banana peel on the path to perfection turned out to be the search for a simple on-off, up-down, spring-driven toggle switch. Sketching rapidly on a pad on his desk, Kraft said that to insure a perfect contact, NASA had ordered a tiny disc mounted on the terminal tips. Silver was specified for the top half of the disc because of its conductivity, and nickel for the bottom half because it could be welded securely.

One day an inspection disclosed that one of the tips had broken off. The switch still worked, but the thought of the tip drifting around weightlessly once the mission had begun and possibly becoming wedged in another switch was enough to start everyone shivering. "What had happened," Kraft said, "was that the disc had been welded upside down and, of course, the silver wouldn't hold. That was easy enough to fix, but then we had to locate every other similar switch in the LM and check it out."

It was this painstaking attention to detail, a fussiness unheard of in preparing more mundane machinery, that had caused the Lunar Module to miss so many production, delivery and test schedules. But it apparently paid off. "Getting the bug ready has been a long, rough road," McDivitt said shortly before the scheduled mission, "but we just finished an altitude chamber run and it came through like no other spacecraft has."

The time had come to see if the strange-looking LM, with antennas sprouting out of its squat body to give it a buglike appearance, measured up to expectations. In space, the astronauts would have to be equally painstaking in the many engineering tests they were expected to conduct. The flight might lack the glamour of going to the moon, but it was an indispensable step in that direction, and those who understood the intricacies of Apollo and moonflight appreciated all that had to be done and what was at stake. "This," observed Russell Schweickart, "is the connoisseur's flight."

The Crew

The job of flying the extremely complex Lunar Module for the first time was entrusted to a former water boiler repairman whose flying career was uncomplicated by romanticism—James McDivitt. A 39-year-old air-force colonel and a veteran of the Gemini 4 mission, McDivitt regarded spaceflight as a professional challenge rather than a personal adventure. Urbane, witty but low-key, McDivitt once remarked that the only reason he got into flying in the first place was to avoid the draft during the Korean War. It was a typical McDivitt put-down of McDivitt that contained only a grain of truth.

McDivitt was born in Chicago, but grew up in Kalamazoo, Michigan, where he went to high school. Afterward he did not know what to do. He worked for a year as a water boiler repairman, then moved to Jackson, Michigan, where he entered Jackson Junior College. It was the Korean War that gave him a purpose. At first he considered merely allowing himself to be drafted, but eventually volunteered for the air force, although he had never flown in a plane. After flight training he wound up in Korea as a fighter pilot, flying 145 combat missions and winning nine decorations.

After the war, McDivitt stayed in the air force and finished college, graduating first in his class at the University of Michigan. He again excelled as a test pilot and, in 1962, was chosen for astronaut training.

The second in command of Apollo 9, David Scott, was born at an airfield, the son of an air-force pilot who became a general, and had been going up in airplanes since the age of 12. A 36-year-old air-force colonel, Scott had had ample experience flying under difficult conditions. He once landed on a golf course when his air-force jet had an engine flameout. Another time he lost an engine flying over the North Sea and just managed to land at a Dutch airfield bordering the water. And an F-104 jet fighter once split apart under him and he ejected to safety. But his most harrowing experience was as the copilot of the Gemini 8 spaceflight in 1966, when the craft tumbled out of control. He and Neil Armstrong, the commander, managed to bring the craft down to an emergency splashdown in the Pacific.

Scott was particularly well suited for the Apollo assignment. After he graduated fifth in his West Point class, he entered the air force, served as a fighter pilot in Europe and, on his return to the United States, entered the Massachusetts Institute of Technology. The thesis he wrote for his master's degree in aeronautics and astronautics concerned interplanetary navigation—ideal training for his chores as the navigator of Apollo 9.

Schweickart, a 33-year-old civilian, had no space experience before Apollo 9. He was born and grew up on a small farm in New Jersey. After earning a degree in aeronautical engineering from MIT, Schweickart spent four years flying jets for the air force. But "Rusty," as he was called because of his red hair and freckles, discovered that he was as interested in science as in aviation. So he returned to MIT to get a master's degree in astronautics, working on a series of scientific projects examining the upper atmosphere by instruments hung from balloons.

The call from NASA interrupted his doctoral studies. Earlier he had not been eligible for consideration as an astronaut because he was too tall—six feet. But the original height limitation had been raised later an inch to six feet, and he was accepted. "I guess being an astronaut was in the back of my mind all the time I was going to graduate school," Rusty Schweickart recalled before Apollo 9.

Countdown and Colds

By late February, all was ready for the launching—all systems were go, as they say in space talk.* The seven-day-long countdown was ticking off without a hitch. In the firing room at Cape Kennedy, the set of numbers flicked silently on the consoles, and as each second passed some-

---

* Once a harried reporter at Cape Kennedy got an urgent telephone call from his literal-minded editor. He wanted to know: "When they say all systems are go, how many systems do they mean?"

one or some computer completed one more check of the 36-story-tall rocket and spacecraft. A sense of tension and excitement was everywhere.

Somehow even Jules Verne failed to think of this arresting touch of drama. His moonbound men, waiting for blast-off, were counted up, and it's not quite the same. The first rocket countdown on record seems to have been devised for no other purpose than its effect. In a footnote to his book *Rockets, Missiles and Men in Space,* Willy Ley, who had worked closely with the German rocket pioneers in the 1920s and 1930s, said that a countdown was apparently first used in the silent film *Frau im Mond* (*Woman in the Moon*) in 1930. At one point, the words "Ten Seconds to Go" flashed on the screen, followed by the numbers "6—5—4—3—2—1—0 Fire." Ley said that much later he asked Fritz Lang, the movie's director, whether he had adapted some military practice, perhaps from World War I. No, Ley said Lang replied, he had thought it up for dramatic purposes.

For Apollo the countdown was a carefully developed procedure enabling launch directors to keep track of all that had to be done to get ready for a lift-off and to ensure that it was all done on schedule, in the proper sequence and before the vehicle was committed to "go." If a malfunction developed, or was suspected, the director would stop the countdown. "T-minus-20 and holding," one might hear on the home TV—meaning that at 20 minutes to takeoff, the clock was stopped. When the problem was cleared up, the count might be resumed or, in many cases, recycled. That is, the countdown clock would be set back several minutes or longer in order to repeat procedures made necessary by the interruption. During the seven days allowed for Apollo countdowns, there were several points where "built-in holds" were scheduled. Often lasting six hours or so, they served two purposes: one, to give launching crews a rest, and two, to give them a time cushion in which to replace any malfunctioning parts that might have been uncovered.

That the countdown also added a new expression to the vocabulary, taught children to count backward almost as soon as they could count forward, and made for moments of unrelieved suspense, was an unavoidable by-product of the Space Age.

Until Apollo 9, the usual reasons for interrupting a countdown were technical problems with the rocket, spacecraft or some tracking station around the world, although Kurt Debus, director at the Cape, enjoyed telling visitors of the early days when they sometimes had to hold the count while a squad of workmen went to the pad to chase away birds.* But about 24 hours before Apollo 9's scheduled lift-off, NASA doctors announced that the flight would have to be postponed for three days. All three of the astronauts had caught the common cold. They had been troubled with sore throats and stuffed noses for several days and were considered too weak and fatigued to go into space. The delay recalled what Melville wrote in *Moby Dick:* "Of all the tools used in the shadow of the moon, men are the most apt to get out of order."

Into Orbit

At 11 A.M. on March 3, the Apollo 9 astronauts finally rocketed into orbit and then successfully executed the first of the complex maneuvers necessary to prove the Lunar Module capable of landing men on the moon. Within 11 minutes after ignition, the three components of the full spacecraft—the Command Module, its rocket-carrying Service Module and the LM—were linked together as one huge package at the top of the Saturn's third stage. The LM was housed in a metal compartment between the rocket and the rest of the spacecraft.

Only three hours after the launching, the Command and Service modules were separated from the LM. Then Scott steered the command ship through a backward flip and gingerly pulsed small rockets to move 50 feet back for a linkup nose-to-nose with the LM, still attached to the Saturn. This was one of the first maneuvers moonbound astronauts must accomplish.

The docking was performed by Scott releasing a latch that allowed the so-called docking probe to push out of the

---

* Merritt Island, where the Saturn 5 pads were situated, is also a bird sanctuary with an abundant and varied population. Before Apollo launchings a helicopter would go around to herd the wildlife to another area. The rumbling of a blast-off would startle them into flight, but the birds would go right back.

Command Module's nose. The probe is a pipe four inches in diameter, surrounded by hinged supports that look like the arms of a scissors-type jack. As the Command Module closed in on the LM, the probe pushed its way into a cone-shaped indentation in the top of the LM. The probe clamped on to the inner rim of the hole. As soon as a "hard lock" between the craft was indicated, Scott opened a valve allowing oxygen from the Command Module to penetrate the tunnel between the two ships. When pressure in the tunnel equaled that in the Command Module, he opened the Command Module hatch and crawled into the tunnel to inspect the docking latches. There were 12 latches, but not all were set to snap shut automatically, lest the shock of this action damage the LM. Scott closed the remaining latches by hand.

Scott checked for proper alignment between the two craft and connected two electrical "umbilicals" from the LM to plugs in the Command Module. These provided electric power to the LM as well as a communications link between the craft. When Scott completed his work, McDivitt radioed to the ground, "Everything came off just right."

Later, Scott flipped a switch to release the linked Command Module and LM from the Saturn rocket stage. From the ground the command went to fire the rocket twice, sending it shooting far out to orbit the sun. The astronauts had maneuvered their craft several thousand feet away by then and could watch the Saturn's spectacular final ignition. Scott radioed: "It's just like a bright star disappearing in the distance."

A Day in Space

Apollo 9's second day in orbit was a time of testing. The astronauts fired the main rocket in the Service Module to shake and roll the linked spacecraft and LM to prove that they were a firm, flyable combination. Nothing dramatic happened, but it was the kind of day Schweickart had in mind when he called this "the connoisseur's flight." It was somewhat typical in that the navigational chores and engine firings performed by the astronauts were standard elements of any Apollo mission. (The following description of the second day of Apollo 9 is adapted from a story by William K. Stevens, a *New York Times* reporter, written at the time to depict a "routine" day in space.)

Sunrise in New York was nearly an hour away at 5:35 A.M. when the three sleeping astronauts, speeding eastward through the blackness over the South Pacific in their 12th trip around the earth, got their first radio call of the day. "Good morning, Apollo 9, Houston," said the voice of Ron Evans, a communicator at the Manned Spacecraft Center. "Good morning, Houston," came the reply from space. "This is Apollo 9."

"Rog," answered Evans. "Loud and clear. Looks like the night was in good shape. We didn't notice any anomalies."

"Very good," said one of the astronauts. "I guess we have to wake up now, huh?"

"Yeah," answered Evans. "It's about that time."

With that exchange began a busy, 12-hour workday for McDivitt, Scott and Schweickart. McDivitt and Scott had elected to sleep in the metal-and-fabric couches from which they carried out many of their daily duties. Schweickart had spent the night in a sleeping bag stretched under his couch.

The interior of their Command Module was a cone-shaped cocoon whose circular base lies under the astronauts' backs as they sleep. The walls of the cone were painted a soft tan as far up as window level, and sky blue above the windows. Ranged around the walls were hundreds of toggle and rotary switches; black circuit-breaker buttons, which, when pulled out, cut off electrical power to a given spacecraft system; meters, valves and electroluminescent displays. McDivitt and Scott, in their couches, awoke to the familiar sight, two feet in front of their eyes, of the ship's main display and control console: a light gray panel shaped roughly like a crescent, arching all the way across the cabin, jammed with switches, gauges, meters, dials, lights, scales, indicators, plotters and miscellaneous displays. Some, like the red "abort" light and most of the 50-odd warning lights, were likely never to be used. Others, like the altimeter in front of McDivitt, would be used only once—after reentry and before splashdown. Still others were monitored routinely, at intervals ranging from one to 24 hours, as a check on the spacecraft's systems.

Just before 6 A.M., Scott vaulted weightlessly into what

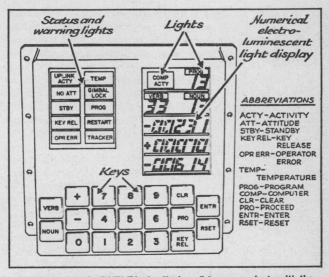

Astronauts use the DSKY (Display Keyboard) to communicate with the
computer. Verb key plus numerals tell computer what action
to take and noun key plus numerals indicate data
to be acted upon.

is called the lower equipment bay (LEB) to bring the command ship fully awake. Like the astronauts, the ship had been drifting somnolently during the night. The LEB, an open space at the astronauts' feet, is the most accessible part of the cabin, the roomiest, and the only one where it is possible to stand up. In it are the navigation system and storage spaces for food, magnetic tapes, film and other supplies.

During the night, the Command Module's guidance system had been all but shut down to conserve power. Now, standing with his back to his crewmates, Scott used switches to feed full power back into the inertial measurement unit that is the heart of the spacecraft's guidance apparatus. Then he brought the ship's computer back to full life after an overnight idling period by briefly punching keys on the DSKY (display keyboard) at his right hand, a mechanism through which astronauts and the computer communicate.

Every job, even that of an astronaut, has its dull routines, and the Apollo 9 crew soon after waking up found itself saddled with one of the dullest: listening through earphones as ground controllers read data needed in case the spacecraft had to return to earth in an emergency, updated information for the flight plan and passed on other necessary data. One astronaut copied the information onto 5-by-8 cards, then read it back to the ground controller as a check on accuracy. It was a job that must be performed often throughout the day. It can get onerous, and astronauts have been known to try to outjockey each other to avoid it.

While one astronaut was copying the data, another was breaking out food packages in the lower equipment bay and was preparing breakfast for the crew. Up to this point, at about 7 A.M., the crewmen had been continually occupied. But it had all been housekeeping and administrative work. The main business of the day was about to begin.

That business was to center on the DSKY, as it most often does in space. This is a flat display area less than a foot square, containing 19 square white keys, five windows where yellow electroluminescent numerals flash and two columns of rectangular lights that tell the astronauts of any trouble in the guidance system and report on the system's status. There is another, identical DSKY on the main display console. Together the two DSKYs constitute, except for the crew, the most commanding presence in the cabin.

Scott used the DSKY in the lower equipment bay to perform the first order of new business, which was to determine how the stable platform of the spacecraft's inertial guidance unit was oriented to the starfield outside. The ship, in effect, rotated about the platform. When the platform was aligned with the stars, the spacecraft's computer had a fixed point of reference for navigation purposes.

As the navigator, Scott was to make two star sightings through the sextant and transfer each sighting into the computer's memory by pushing a "mark" button when he had the proper star lined up. The entire process was controlled by a special program stored in the computer's memory. To start the process, Scott instructed the computer through the DSKY to initiate Program 51. This he accomplished by pressing keys marked "Verb 37 Enter, 51 Enter." (In computer code, "Verb" means an action to be taken. "Thirty-seven" is the specific action—in this case, "Change Program." "Enter" means proceed with the action. "Fifty-

one" is the program desired—the star-sighting process.) On
the DSKY's electroluminescent numeral display appeared
"V 50 N 24; 00015." Rough translation: "Please sight on a
star." The display was flashing, indicating that a response
from the astronaut was required.

Scott was ready to comply. He punched the "Pro" button
on the DSKY, meaning "Proceed." This sent the computer
into a preparatory routine; and when it was finished it came
back with a flashing "V 51," meaning "Please mark."
Scott pushed the "mark" button, and the star sighting was
recorded.

The process was repeated with another star, and the
computerized guidance system was now equipped to take
part in the day's pièce de résistance: a firing of the Command Ship's main engine to see how well the Command
Ship and the Lunar Module performed when linked together under acceleration.

To that end, in deliberate succession, ground controllers
radioed data, for use by the computer, about how fast the
craft was going and about its position; about how long the
main engine was to fire, and in what direction its thrust was
to be. The controllers sent the same data to the astronauts,
who again, laboriously, copied it down for use in double-
checking the computer.

As time for the 9:12 A.M. firing of the engine drew
near, the crewmen were busy checking the instruments that
would be involved, performing checklists of tasks, stowing
items that might crash about during the sudden acceleration and donning their protective spacesuits and helmets.

At 8:35 A.M., one of the astronauts punched into the
DSKY, "Verb 37 Enter, 40 Enter," starting the programed sequence that culminated in the automatic, com-
puter-directed engine firing.

The immediate postfiring period was described by one
Houston flight planner as "a lull." But Walter Cunningham,
an Apollo 7 astronaut following Apollo 9's flight, said that
"the lulls aren't what they seem." "In reality," he said, "it
may take you two hours to get the food reconstituted and
eaten. So I imagine those guys will still be eating breakfast
after the burn."

The remainder of Apollo 9's day consisted essentially of
two repeat performances of what had gone on up through
the engine firing, interspersed with another meal and

periodic housekeeping chores. At 5:30 P.M. the three astronauts settled down for the night and the three much busier days ahead of them.

The Human Spacecraft

On the third day, March 5, the LM passed its first manned-flight test.

Wrapped in oxygen hose, hampered by bulky, partly inflated spacesuits, the three astronauts removed the Command Module hatch and reached through the connecting tunnel to open the Lunar Module hatch. Then Schweickart and later McDivitt crawled through the tunnel (38 inches long and 32 inches wide) into the LM. The two men spent nearly nine hours there, while Scott remained to pilot the Command Ship. McDivitt and Schweickart switched on nearly all of the LM's instruments, extended its four spindly legs, sampled its computer data to make sure it was providing accurate information, and ignited the descent engine for a six-minute firing. They also transmitted a short telecast of their activities inside the LM cockpit.

The LM was declared in good shape—which was more than could be said for Schweickart at the time. Twice during these activities he was hit by severe attacks of nausea, perhaps a form of motion sickness. For a time, flight controllers canceled plans to have Schweickart take a two-hour space "walk"—called extravehicular activity, or EVA. Though not a high-priority part of the mission, EVA was included to give the astronauts training in the emergency procedure for moving from one vehicle to another if something should block passage through the tunnel or if, on the return from the moon, the two ships are unable to link up when they rendezvous. It would also provide the first real test of the oxygen-supplying backpack men would use while walking on the moon's surface.

When the astronauts awoke the next day—March 6, their fourth day in orbit—Schweickart had recovered from his nausea and was not to be denied some kind of excursion in open space.

McDivitt and Schweickart, who had returned to the Command Module for the night, went back through the tunnel into the LM. All three astronauts were fully suited, breathing oxygen piped directly into their suits. Both cabins were vented of oxygen so that as complete a vacuum existed inside as outside in space. Then hatches on both vehicles were opened.

Scott stuck his head and shoulders out of the Command Module hatch to retrieve some experiments that had been attached outside. As Apollo 9 soared over the Pacific Ocean at sunrise, Schweickart stepped out on the LM's porchlike platform. He spent most of his 40-minute EVA with his booted feet securely planted in a pair of "golden slippers" attached to the porch. These were hard-rubber restraints that were covered with gold paint as protection against the sun's intense rays. Being restrained in this manner—a lesson learned from the Gemini EVAs—Schweickart could use both hands to operate cameras. "That's what you call a view from the top of the stairs—one stair, that is," Schweickart commented.

Even though the EVA was abbreviated, it was long enough to prove the effectiveness of the $2 million worth of spacesuit and portable life-support system that men would use on the moon. For when Schweickart was outside, he was the world's first self-contained human satellite. It was the first time an astronaut had left his orbiting craft and depended solely on a backpack for his breathing oxygen, air conditioning and communications rather than on umbilical hoses leading from the vehicle.

Schweickart's backpack—called the portable life-support system, or PLSS—weighed 84 pounds and was 26 inches high, 18 inches wide and 10 inches deep. Inside it was a replaceable silver zinc battery—about the size of a pocket transistor radio—that produced power for the PLSS systems.

The oxygen supply system provided breathing and controlled pressurization in the suit from the pound of oxygen stored in a tank. The ventilating circuit forced oxygen through the suit and backpack to remove carbon dioxide and contaminants and to cool and remove perspiration. The moisture was then removed by a water separator. The water-transport loop cooled the astronaut by removing his heat of metabolism and any heat leaking into the suit from outside. A pump drove the water through a network of tubes that ranged in thickness from ¼ to 1⁄16 inch. The feed-water loop supplied expendable water, which was stored in a bladder, to cool the system. The water was

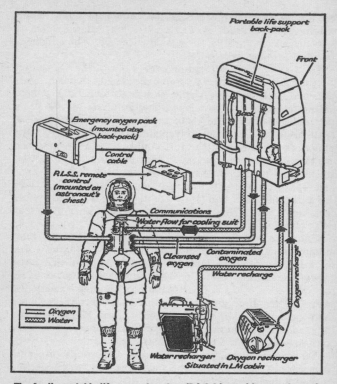

The Apollo portable life support system (P.L.S.S.), used by an astronaut outside the spaceship, makes him a self-contained human satellite with his own oxygen, air conditioning and communications.

forced outside the suit, turned to ice and vaporized. A communications system included two radios, biomedical data transmitters and warning systems, should something go wrong. Monitors transmitted to the ground data on oxygen supplies, water temperatures, suit pressures and the battery current. One radio channel transmitted an electro-cardiogram signal.

A second backpack, weighing only 41 pounds, was called the oxygen purge system and was designed for use should the PLSS fail. The PLSS was designed to run unrecharged for four hours, while the second pack could provide enough oxygen for 30 minutes.

Rendezvous

After Schweickart's EVA, the astronauts shut the hatches and returned to the Command Module to rest up for their fifth and most critical day in orbit—the day of rendezvous. This would demonstrate how well the astronauts who landed on the moon in the LM could, after their brief inspection of the lunar surface, find and rejoin the moon-orbiting Command Module for the trip back to earth.

It was a spectacular display of multiple navigation that lasted for about 6½ hours. McDivitt and Schweickart went back into the LM and, when all was ready on Apollo 9's 59th revolution of earth, Scott tripped a switch in the Command Module cockpit to release the 12 latches holding the two ships together at the tunnel.

"Spider" was then free of "Gumdrop." In their radio communications between the two ships the astronauts referred to the LM as Spider, because of its buglike appearance, and to the Command Module as Gumdrop, because the cone-shaped vehicle was delivered from the factory to the Cape in a blue, plastic, candylike wrapping.

After making sure that Spider was functioning perfectly, McDivitt and Schweickart fired small LM rockets to pull their ship away from Scott and Gumdrop. For an hour they drifted within a few miles of Gumdrop, checking out their rendezvous radar and guidance systems. Then McDivitt fired up the LM's descent rocket to climb into a 156-mile-high orbit, 13 miles above the Command Module. "It got a little rough and shaky as I was throttling up," McDivitt reported. The LM continued on its course for nearly two hours while the two craft circled the earth, their maximum distance from each other reaching 50 miles.

Then when Spider got back within sight of Gumdrop, McDivitt refired the descent rocket. Shortly afterward he triggered explosive bolts to jettison the LM's lower stage with its four legs, thus exposing the nozzle of their ascent rocket. When the LM, flying higher in an "outside track," fell behind about 100 miles, McDivitt fired the ascent engine in a simulated takeoff from the moon. In a series of subsequent firings, calculated by the LM's computer from measurements taken by the craft's radar, McDivitt and Schweickart steadily closed the gap between themselves and Gumdrop. If they had failed to find Gumdrop, the two men in Spider would have been stranded in orbit to die when their oxygen supply ran out.

But their radar and rockets worked flawlessly. When the squat, buglike LM hove into view, Scott radioed greetings: "You're the biggest, friendliest, funniest-looking spider I've ever seen."

Finally, coached by Scott, McDivitt slowly brought the LM up to the mother ship for the redocking. After McDivitt and Schweickart squeezed back through the tunnel into the Command Module, the astronauts released the LM and, on a radioed command from the ground, the ascent engine was refired until all its fuel was depleted. Both the descent and ascent stages of the LM were now only so much space junk which would eventually burn up as they dipped back into earth's atmosphere in several weeks.

After completing the crucial rendezvous, the Apollo 9 astronauts were left alone in orbit with fully 97 percent of their mission objectives already completed, but with five more days of flight. They made good use of the time, practicing navigation and tracking skills and conducting photographic experiments designed to study the earth's resources. On March 13, the three astronauts brought their spacecraft to a smooth and accurate splashdown in the Atlantic Ocean north of Puerto Rico. It hit the water less than a mile away from its target point and only three miles north of the waiting recovery ship, the U.S.S. *Guadalcanal*. Television cameras on the helicopter carrier were able to show the worldwide audience the last moments of the parachute-aided descent and all of the recovery operations by helicopter and frogmen. The men were safe and healthy after ten successful days in space.

Onward

Thus, in earth-orbital rehearsal, was accomplished the most critical phase of the lunar mission—the rendezvous between the LM, soon to return from the moon's surface, and the

Command Module, waiting in lunar orbit. Apollo 9 demonstrated that the LM was spaceworthy. It demonstrated the successful design of the docking system that enabled two craft to join and lock together in space, and the tunnel connection that allowed crewmen to transfer back and forth without venturing out into raw space. Finally, it tested for the first time in space the life-support backpack that would enable men to step onto the moon without any connecting link with their spacecraft. Apollo 9 did all these things and, to the surprise of nearly everyone, did them almost without a hitch.

At a news conference in Houston after the splashdown, George Mueller, NASA's associate administrator for manned spaceflight, smiled broadly when he declared: "Apollo 9 was as successful a flight as any of us could ever wish for, as well as being as successful as any of us have ever seen."

Two other developments occurred during Apollo 9's flight to assure the moon project's continued onward thrust. The first was that on March 5, Thomas O. Paine, 47, was given clear title as NASA's administrator. As acting administrator since James Webb's resignation in October, 1968, Paine had impressed nearly everyone with whom he worked —scientists, industrial contractors and engineers at NASA's many operations centers and laboratories around the country. These experts found him sharp and quick in his thinking, cogent in his arguments and dedicated to space exploration, especially the fulfillment of the Apollo goal. Though a nominal Democrat, Paine was appointed NASA administrator by the new Republican president, Richard M. Nixon, who had taken office in January.

The second development was that the door in the Vehicle Assembly Building at the Cape opened once again and yet another Saturn 5 moved toward the launching pad. This was the rocket and spacecraft for Apollo 10, scheduled as the final test before the landing itself.

Chapter XVI

Apollo 10: "The Forgiving Flight"

Without question the most popular comic-strip characters of the 1960s were a round-faced perpetually puzzled little boy named Charlie Brown and his rambunctious beagle Snoopy. The world Charlie Brown viewed with fragile wonder in the "Peanuts" comic strip seemed always too much for him, more complicated than it should be, hard and friendless, frustrating. Try as he might, he could never win a baseball game, or screw up the courage to talk to the cute little red-haired girl. But try he always did, until it hurt—especially those kindred spirits of the bewildering decade who followed his daily efforts to cope. Snoopy, blessed with more imagination and spunk, simply ignored the fate that decreed him a dog and went merrily along in a dream world atop his house. His flights of fancy carried him, after the fashion of Walter Mitty, into World War I aerial combat with the Red Baron and on his own moon journey, beating "The Americans, the Russians, and that stupid cat next door." "My favorite gag of the whole series was on the return trip," Charles M. Schulz, Snoopy's creator, said in an interview with Steven V. Roberts of the *Times*. "Snoopy turns around with that big grin of his and says: 'You can tell we're on the way back because we're facing the other way.' "

Snoopy was no stranger to Apollo. In 1967, after the fire, NASA officials started a safety and morale-building program with contractors and picked the irrepressible pooch as its symbol. "We thought Snoopy as an astronaut would be perfect," a NASA official explained. "He happened to be the only dog with flight experience and was thus eligible to be a mascot for the astronauts." Snoopy started appearing on posters around NASA dressed in full space regalia, complete with a bubble-type helmet. Individuals who performed outstanding work on Apollo received silver Snoopy lapel pins.

By May of 1969 Charlie Brown and Snoopy were set to achieve new heights. Like the Walt Disney and Milton Caniff characters who decorated the plane fuselages of World War II, Charles Schulz's two most famous creations were designated as the code names for the Command and Lunar Modules of Apollo 10. "Using Charlie Brown on the spaceflight is almost tempting fate, isn't it?" commented Schulz, his voice quiet and merry as he played with the idea. "We once thought about putting out a Charlie Brown baseball glove, but what kid would want one? He couldn't catch a thing with it."

Thomas P. Stafford, John W. Young and Eugene A. Cernan, the men preparing to fly Apollo 10, were willing to tempt fate. They were all veterans of Gemini flights and had been training more than a year for their Apollo mission. They had been briefed by the crews of Apollos 7, 8 and 9. They knew what they had to do; fly around the moon 31 times, swoop down in the Lunar Module to nine miles above the intended landing site, and then return to earth. Their planned eight-day mission was billed as the dress rehearsal for the actual landing flight, for they would practice everything except the actual touchdown and lift-off from the moon. They would, as one reporter said, do everything but make history.

Near-Miss

In that sense, fate had already taken a disappointing turn for the Apollo 10 astronauts. When they were named as a crew the year before, they were assigned a Lunar Module to be fully equipped and programed for a landing on the moon. Development delays with the vehicle, however, forced NASA to make the schedule changes that led to the Apollo 8 flight around the moon without the LM and shifted the landable landing craft from Apollo 10 to Apollo 11. Even so, up until late March, there was still an outside chance that the Apollo 10 crew might make the first landing. Because of the almost flawless Apollos 8 and 9 flights, some consideration was given to skipping another moon-orbiting mission and getting on with the landing itself. But the Apollo program director, General Samuel Phillips, finally ruled that out.

Phillips had emerged, through the years of Apollo's preparations, as the project's driving force and key decision maker. As he explained it to me, "I'm at the level which knows all the things you have to know to make a major decision. Below the program director, there isn't anyone who has the whole picture. Above the program director, the men have so many other responsibilities."

Through daily meetings, countless telephone conversations and visits to space centers and contractors' plants that kept him on the road 75 percent of the time, Phillips was able to assess the flight readiness of equipment and crews and to identify potential problems early enough so that plans could be changed. It was Phillips who bore the primary responsibility for the decision to fly Apollo 8 around the moon on the first manned test of the Saturn 5. This kept the program from losing valuable time in waiting for the Lunar Module to be made ready. Phillips told me afterward that the Apollo 8 decision was his most difficult one because it fell in the class of choices "where you can't be wrong."

To be around Phillips was to be impressed with his tough-minded thoroughness. He was a tall, usually taciturn air-force lieutenant general on loan to NASA, and although he wore civilian clothes and worked in his shirt-sleeves, he was unmistakably the military man when he squared his shoulders, set his firm jaw and fixed his blue gray eyes on the person to whom he was talking. "Sam is not a man to speak lightly," an associate said. "At our meetings he sits quietly, takes detailed notes and asks extremely tough questions. He rarely loses his temper, but he has a way of quietly and firmly letting someone know of his displeasure." At a small dinner party before the Apollo 10 launching, Wernher von Braun was speaking of the incredible way all the Saturn 5s thus far had been launched within milliseconds of the scheduled time. Von Braun singled out General Phillips as the man to whom the greatest credit should go for pulling together the many pieces of Apollo and making them work—on time.

After listening to the Apollo 9 crew's debriefings and sounding out the flight controllers at Houston, Phillips made up his mind that one more rehearsal flight was ad-

visable. It was too great a risk to move directly to the
landing after only one flight around the moon and only one
manned test of the Lunar Module. "We needed the Apollo
10 mission," Phillips told me, "to mature the total system
of equipment and men. We especially needed more experience with the Lunar Module."

Phillips set the launching for May 18. Scheduled highlights of the mission included:

- Flying the Lunar Module to within nine miles—
  50,000 feet—of the moon, the closest man had ever
  gone to another celestial body.
- Accomplishing a rendezvous and linkup between the
  Lunar Module (Snoopy) and the Command Module
  (Charlie Brown) at about 70 miles from the moon
  and some 230,000 miles from earth.
- Determining what effect the moon's weak and somewhat irregular gravitational field, owing to the so-
  called mascons, had on the accuracy of flight
  maneuvers.
- Photographing Apollo 11's intended landing site on
  the Sea of Tranquillity.
- Circling the moon's equator 31 times in 2½ days,
  three times longer than the ten orbits of Apollo 8.
- Transmitting the first color television pictures from
  deep space.

"I'd be fooling if I said I wasn't disappointed at not
making the landing," Cernan said at the time. "If I didn't
think our mission was necessary, I would have fought
harder to make the first landing."

Apollo 10, therefore, was planned as the "forgiving
flight," in the words of George Low, the spacecraft manager. By that Low meant that Apollo 10's limited mission
around the moon would presumably bring it home with its
tanks half full, leaving a wide margin for error or oversight. Engineers would be forgiven any less than catastrophic mistakes and could actually learn from them.

The Men for 10

"It's definitely the riskiest of any mission put together,"
Thomas Patten Stafford, the Apollo 10 commander, said

before lift-off. "The only one that could be more risky is the actual landing itself."

The appearance of the big, baldish, 175-pound astronaut with his slow Oklahoma drawl seemed to mark Stafford more as a schoolteacher, accountant or banker than as one of the best qualified of the corps of astronauts. But his knowledge of engineering, mathematics and computers set him above the usual run of test pilots. The 38-year-old Stafford was a colonel in the air force at the time of Apollo 10, although he attended the U.S. Naval Academy, where he stood near the top of his class of 1952. He helped write two basic textbooks for test pilots and attended the Harvard Graduate School of Business Administration—for three days.

After trying and failing to become one of the original seven astronauts because, at six feet, he was one inch too tall according to the initial standards, Stafford was admitted to Harvard in 1962. But after those three days, he was notified that he had been selected as one of the second group of astronauts (the acceptance standards had been changed, making him eligible). He left graduate school without hesitation. His rise through the astronauts' ranks was rapid. After being a backup pilot for Gemini 3, he was a member of the Gemini 6 crew with Wally Schirra and later commanded Gemini 9.

Stafford's fellow pilot in the Lunar Module was Eugene Andrew Cernan, a navy commander who had flown with Stafford on Gemini 9. He was a slim, 35-year-old six-footer with short salt-and-pepper hair. He took his job seriously, but not himself or his status as an astronaut. Once a carload of tourists drove up in front of Cernan's home near the Manned Spacecraft Center and asked him where some of the astronauts lived. He was dressed in a T-shirt and ragged work pants, tending his yard and garden. Cernan scratched his head and told the tourists that "I think some live there at the end of the block . . ." He strolled into his house, grinning broadly.

Cernan was born in a Chicago suburb and received a degree in electrical engineering from Purdue, where "my appetite for flight was whetted." After Purdue, he entered the navy through the reserve officer training program and immediately signed on for flight school. He became a crack test pilot, and was selected as one of the third group of astronauts in 1963.

The man who would remain in the Command Module while Stafford and Cernan scouted the moon up close was John Watts Young, a 38-year-old navy commander. Young was something of a loner who could be both witty and aloof, mechanically minded and adventuresome, yet hating the outdoors. "John is an amusing bundle of such contradictions," his wife once remarked.

During his initial spaceflight aboard Gemini 3, it was the usually solemn-faced John Young who pulled a kosher corned-beef sandwich from his pocket and offered a bite to his startled commander, Gus Grissom. The aftermath of the prank saw the remains of the sandwich enshrined in a plastic case atop Young's desk beside an official reprimand from unsmiling directors of the space agency. Yet hardly a year after the incident he was back in space as the leader of the Gemini 10 mission, which set a world altitude record at the time of 475 miles.

The green-eyed, 165-pound Young breezed through high school in Orlando, Florida, with straight As, and earned a degree in aeronautical engineering from the Georgia Institute of Technology in 1952. It was during his college vacation that he first saw the area of the future Cape Kennedy while helping to survey a rocket-launching site. As a navy pilot, he set two world records for high-performance jet aircraft in 1962, the same year he was chosen for the astronaut corps.

As the Apollo 10 launching day approached, the three astronauts tapered off their training so as not to fall victim to the fatigue and colds that hit the Apollo 9 crew. They were pronounced in good health. And so was the technological star of Apollo 10—the Lunar Module called Snoopy. Two new features aboard the Apollo 10 LM that had not flown before were the landing radar system and the S-band steerable antenna for communications at lunar distances. The LM computer was also different, being programed for an eight-hour flight around the moon and two dips down to within nine miles of the surface. The computer was the primary navigator. "These men," explained Joseph Gavin, the Grumman engineer in charge of the LM development, "are really computer experts playing numbers into their computer keyboards, rather than flying the spacecraft in the conventional sense of airplanes."

Asked if the LM was in perfect condition and ready for blast-off, Gavin replied, "I shouldn't say so, but I think it is." He crossed his fingers and looked out toward Launching Pad 39-B.

"I Wish We Could Stay"

Another perfect launching sent the three astronauts on their moonward course early on a Sunday afternoon. After circling the earth 1½ times, the Saturn 5's third-stage engine refired for five minutes to propel the spaceship away from its low earth orbit at an initial speed of about 24,200 miles an hour. "We're on the way!" radioed Stafford as the rocket lighted the dawn over the Pacific Ocean.

In a maneuver identical to the one on Apollo 9, the astronauts separated their Command Module from the rest of the "stack," did a back flip and moved its nose in for a docking with the Lunar Module, nestled on top of the Saturn 5 third stage. Then, while their color television camera was grinding away, the astronauts pulled the attached Lunar Module out of its nest and watched the Saturn 5 recede in the black void of space.

For three days Apollo 10 coasted on its course, losing speed until, within 38,000 miles of the moon, they were pulled ever faster toward the satellite. It was a relaxed time. Some of the problems of the earlier flights—the colds and nausea—were missing. The astronauts' only complaints were too much chlorine in the water (owing to an erroneous valve setting), hydrogen bubbles in the water causing stomach gas, and some drifting snowlike particles of glass fiber from a rip in the Command Module's hatch covering.

At 4:45 P.M., eastern daylight time, on May 21, Apollo 10's Service Module rocket fired to slow down the vehicle and send the astronauts into orbit of the moon. The astronauts then turned on their color television camera to give the people back home a 29-minute show of what it was like to sail over such pockmarked places as the Sea of Fertility, the Sea of Crises, Langrenus Crater and the Sea of Tranquillity. On the southwestern part of the Sea of Tranquillity, near the crater Moltke, lay the smooth plain that was to be the landing site for Apollo 11.

Early the next morning, Cernan and Stafford crawled through the connecting tunnel into Snoopy leaving Young behind in Charlie Brown. After checking all instruments to make sure they had survived the 250,000-mile journey, the two astronauts separated the LM, flew formation for a while and then, firing the LM's descent rocket, streaked down through the lunar sky at a speed of 3,700 miles an hour. "We're right there! We're right over it! I'm telling you, we are low, we're close, babe. This is it!" Cernan exclaimed as Snoopy swooped to within 47,000 feet of the moon's surface—not much higher than the altitude at which commercial jets fly over the earth.

There were several reasons for choosing such an altitude for Snoopy's scouting expedition. The altitude of about 50,000 feet was the last point where, on an actual landing mission, the crew could decide to refire the rocket for the final seven minutes to touchdown. If there were problems, they could fire the ascent rocket at the same point to head back for a rendezvous with the Command Module, keeping watch in its orbit 69 miles overhead. That altitude of about 50,000 feet was also where the LM's landing radar, which must control the final descent, was designed to become effective, bouncing signals off the lunar surface to determine the LM's altitude and speed. A third reason was that, at 50,000 feet, it was still possible for the Command Module to plunge down to rescue the LM, should the LM's rockets fail. It was possible for vehicles to draw that close to the moon without crashing into it because of the absence of any detectable atmosphere. It is atmospheric drag, as well as gravity, that pulls a satellite out of a low orbit and down to the body it is orbiting.

On Snoopy's first low pass, Stafford looked out the window and described Tranquillity's landing site as "very smooth, like wet clay, like a dry riverbed in New Mexico or Arizona." He was able to pick out more details—boulders, small craters and deep cracks—on the lunar surface than he could have done looking down at earth from that altitude. The moon's lack of any atmosphere meant there were no clouds or haze, no gases acting as distorting filters. The sight gave the astronauts a moment of understandable wistfulness. Said Stafford: "I just wish we could stay."

But that was not in the flight plan for Apollo 10. Instead, the two astronauts braced themselves for the throttlable descent engine to refire in the direction of the moon, thus boosting them up and out into a wide orbit taking them 220 miles above the surface at the highest point on the moon's far side. The orbit brought the Lunar Module back around the moon and down over the Sea of Tranquillity, for its second close pass.

Just before the second flyover, the astronauts had their scariest experience. Explosive bolts were automatically triggered to jettison the LM's descent stage with its four spindly legs. At that moment the roundish upper stage went into a fast spin, pitching up and down. "Son of a bitch," shouted Cernan. "Something is wrong with the gyro"—the guidance system. Taking over the hand controls, Stafford steadied the spacecraft after about a minute. "I don't know what the hell that was, baby," Cernan told ground controllers, "but that was something. I thought we were wobbling all over the skies."

What had happened, ground engineers later concluded, was that a control switch had been left in the wrong position. This was presumably done by technicians who had prepared the LM before launching, and had not been spotted by Stafford when he inspected the instrument panel.

By the time Snoopy and its pilots had settled down, they were again down within nine miles of the moon. "I'll tell you," Cernan commented, "We're down there where we can touch the top of some of the hills." At this time, the LM's ascent engine fired 15 seconds to boost the two astronauts back toward their rendezvous with the Command Module, following a curving upward course similar to the one astronauts taking off from the lunar surface would follow. "Snoopy and Charlie Brown are hugging each other," one astronaut shouted as the two craft were joined.

After Stafford and Cernan crawled back into Charlie Brown, they closed the hatches and jettisoned the unmanned Lunar Module's upper stage. Its rocket was fired until all the fuel was consumed, sending the craft out into a wide orbit of the moon. Said Cernan: "That little Snoopy was a real winner."

To hear Stafford describe it, a ride in the Lunar Module's

upper stage was quite an experience. When the smaller rockets were fired for the closing phases of the rendezvous, he said the thin aluminum alloy walls of the vehicle shook and rattled. "If you wanted an LM simulated ride," Stafford told ground controllers, "let your kids get a big metal bowl on your head and beat on it with spoons."

The Apollo 10 astronauts continued orbiting the moon for another full day. From their orbit, ranging from altitudes of 63 to 75 miles, they observed craters "looking like live impacts" and craters that looked "like volcanoes"—thus giving some ammunition to both sides of the scientific argument about the origin of the craters. "There's one on the back side," Stafford said, "that if it was in a different setting you would call it Mount Fujiyama." Later, Stafford reported seeing a crater whose outside slope was littered with "so many big, black boulders" that it looked "like a forest of pine trees."

A Clean Shave

With a powerful push from their spacecraft rocket, the Apollo 10 astronauts broke out of moon orbit early Saturday morning, May 24, and headed for earth with ever-increasing speed. Their aim was true. Their spacecraft was still performing smoothly. They were tired but in good spirits after circling the moon 31 times.

As Apollo 10 drew away from the moon, the astronauts transmitted some of the most spectacular television pictures of the mission. During the 52-minute telecast, the spaceship seemed to be climbing straight out from the center of the lighted part of the moon until it was possible to see the moon as a full sphere. "As we move away," Stafford radioed, "the basic moon looks tan to us." And it was set "against the blackest black you ever saw," Stafford added. The crater Tsiolkovski, one of the largest landmarks on the far side of the moon, was clearly visible on the screen. There were also shots of the whole of Smyth's Sea and the larger Sea of Tranquillity. Stafford made the interesting observation that the Sea of Serenity seemed to be a lighter brown than the more chocolate-colored Sea of Tranquillity. This may have been the result of different lighting conditions.

Looking back on the scarred face of the moon, Cernan

remarked: "I tell you, this satellite of ours—this moon of ours—had a rough beginning somewhere back there."

During Apollo 10's 54½-hour return voyage, the astronauts relaxed most of the way and even took time out on Sunday to achieve another "space first." All three men shaved. "We're in the process now of commencing scientific experiment Sugar Hotel Alpha Victor Echo," Young radioed. "And it's going to be conducted like all normal human beings do it."

Shaving in space under weightless conditions turned out to be easier than anyone had imagined. The three men simply took out a tube of brushless shaving cream, a safety razor and towels and attacked their seven-day stubble.

For several years NASA had experimented with electric shavers that included suction attachments to collect the loose bristles, keeping them from floating around freely in the cabin and possibly clogging up some of the electronics. These complicated devices, on which NASA was said to have invested $5,000 in research, were apparently never satisfactory.

With plain, off-the-shelf (from a Cape Kennedy drugstore) brushless shaving cream, the astronauts had no problems. After each stroke with the razor, they wiped the razor clean with a small paper towel. The whiskers adhered to the cream, never floating out into the cabin. "It was a three-way huge success," Young reported.

On May 26, the three astronauts stowed away all the loose gear in the cabin, got "updates" for their computer and braced themselves for the blazing ride back through earth atmosphere to splashdown. They had required only one mid-course firing of the spacecraft's maneuvering rockets to sharpen their aim on earth.

As they did, the astronauts beamed one final telecast of their trip. More than any previous astronauts, they had sought at nearly every point in their flight to give the folks back home a vicarious feeling of participation. Their 19th telecast focused on the blue and white sphere that was earth. "Through this endeavor," Stafford said of the telecast, "we hope that we have made you and millions of people of the world more a part of the history that's being made in our day and age."

Then Apollo 10 went through the same sequence of

maneuvers that Apollo 8 had followed to get back to earth.
The Service Module was jettisoned. An automatic command from the computer fired maneuvering rockets to
point the blunt end of the cone-shaped Command Module
toward the earth at an angle. This edge would bear the
brunt of the spacecraft's fiery reentry. There was the communications blackout from the heat, the slowing down of
the vehicle from the atmospheric friction. Finally, the parachutes unfurled and, as dawn came to the South Pacific,
Apollo 10 dropped into the water within sight of the U.S.S.
Princeton, a helicopter carrier standing by for the recovery.
It was 12:52 P.M., EDT.

"Tell the medical officer to take a couple of aspirins and
relax," one of the astronauts radioed from the bobbing
spacecraft to the hovering helicopter. "We feel great."

"We Will Go"

Within an hour after the astronauts stepped on the *Princeton,* Thomas Paine, NASA's administrator, left the control
room and walked over to the Manned Spacecraft Center's
auditorium to read a statement to reporters. It was the
official go-ahead signal for the Apollo 11 astronauts—
Armstrong, Aldrin and Collins—to press on with their
preparations for the moon-landing attempt.

"Eight years ago yesterday," Paine said, "the United
States made the decision to land men on the moon and
return them safely by the end of the decade. Today, this
moment, with the Apollo 10 crew safely aboard the U.S.S.
Princeton, we know we can go to the moon. We will go to
the moon. Tom Stafford, John Young and Gene Cernan
have given us the final confidence to make this bold step."

About the only question left, Paine said, was the timing.
July 16 had already been set as the tentative launching
date. A preliminary flight plan already called for the landing to occur July 20. But the Apollo commander, Neil
Armstrong, had expressed some concern a few days before
over the tightness of the crew's training schedule. If there
were any slip in critical training procedures, Armstrong
told NASA officials that he would like to postpone the
flight a month to a time in August when the lighting con-

ditions over the landing site would again be favorable. Paine had assured Armstrong that if his crew needed more time, they could have it.

"We have no false pride about these specific dates," Paine told reporters at the Apollo 10 splashdown news conference. "We will not hesitate to postpone this mission if we feel we are not ready in all respects."

But Paine, Phillips and Low said that a delay was unlikely. On May 20 the Apollo 11 rocket and spacecraft were trundled to the launching pad at Cape Kennedy. They were fully assembled, and the pad check-outs were proceeding on schedule. Moreover, nothing in the performance of Apollo 10 indicated the need for any serious changes in the equipment, launching procedures or flight plan. "Today we see no obstacles on the path to the moon," Paine concluded.

Apollo 10's success had two other consequences. London bookmakers announced their refusal to make any more bets against man's being able to land on the moon in the decade. And, more significantly, scientists in the Soviet Union told the Russian people what most of them already assumed—that the United States would make the first attempt to land men on the moon. Their concession appeared in Moscow newspapers. Said Leonid Sedov, writing in *Pravda:* "Mankind highly appreciates the achievement of the American specialists and the daring astronauts." One man in Red Square, when asked his opinion of Apollo 10, gave the Russian equivalent of "Your boys showed ours a thing or two."

Thus the great moon race was over, as far as the Russians were concerned, but Armstrong and Aldrin, along with Collins, still had to cross the finish line.

Chapter XVII

Men For The Moon

"When ships to sail the void between the stars have been invented," said Johannes Kepler, the German astronomer who more than three centuries ago helped to explain how ships would one day sail that void, "there will also be men who come forward to sail those ships."

The ships had been invented, and of the 3.5 billion people in the world, the 202 million Americans and the 50 active astronauts, three men were selected to make the first voyage to land on the moon. Their colleagues of Apollos 7, 8, 9 and 10 had demonstrated the spaceworthiness of the vehicles, the speed and precision of the worldwide network for tracking and control and the soundness of the flight techniques for lift-off, rendezvous and reentry.

The three men for the moon assignment had worked together as a team for more than a year. First, as the backup pilots for Apollo 8, they learned through simulations and long textbook study the ways of a Saturn 5 in flight, of a Command Module on the way to the moon and of a fiery reentry to splashdown. Then, since January, 1969, when they were chosen for Apollo 11, they had practiced for weeks on end at the controls of the Command Module and the Lunar Module. The two men who would actually land on the moon spent hours learning geology and photography and studying lunar maps to be able to make the most of the short time they would spend on the moon's surface.

The three men for the moon journey were Neil Alden Armstrong, a 38-year-old civilian; Edwin Eugene Aldrin, Jr., a 39-year-old Air Force colonel, and Michael Collins, a 38-year-old Air Force lieutenant colonel. Two of them, Armstrong and Aldrin, would actually plant their booted feet on the moon.

What manner of men were these three? What set them apart from the rest of their fellow Americans? In what ways were they very much akin to so many men who would never go to the moon? Some answers to these questions emerged in the weeks before the Apollo 11 flight from talks with the three astronauts, visits to their home towns, and interviews with their parents, brothers and sisters and people who had worked with them.* Following are word portraits of the three who were destined for a high place in history.

Armstrong

Neil Alden Armstrong walked into the room, a conservative plaid sports jacket draping his small-boned 5-foot, 11-inch frame, his blond hair parted on the left and slicked down like a Sunday school boy's. He was about to shake hands and talk with a stranger, and his odd sort of lopsided grin suggested he was trying hard to meet the stranger halfway.

That did not appear to come easily to the commander of the United States' first lunar landing mission, the man chosen to set first foot on the moon. He was surrounded by protective walls of shyness, modesty and self-possession. During the half-hour chat in Houston—his part in a round-robin series of conversations with the men of Apollo 11—Armstrong comported himself formally, often pausing a long time before delivering a sparse answer to a question.

On the basis of such an encounter, you would never know that he was once a champion beer chug-a-lugger in the nightspots around Edwards Air Force Base, on the Mojave Desert of California, where he once flew the X–15 to the fringes of space and earned a reputation as one of the world's best test pilots; that he was a shrewd player of the stock market; that among close friends, he was one of the most cherished of companions; that he had lived for one thing—flying—ever since he was a precocious, introverted little boy who often would rather stretch out with a book on the living-room floor than play with other children.

---

* Conducted by William K. Stevens of *The New York Times*.

Armstrong was born in the living room of his grand-parents' farmhouse six miles southwest of Wapakoneta (population about 7,000), a place whose open manners, tree-lined streets and abundance of old-fashioned two-story clapboard houses made it almost a model of small-town mid-America. It was an area of northwestern Ohio settled in the late 1860s by families of solid German farmers and merchants fleeing the draft under Bismarck's blood-and-iron policy. In succeeding years these immigrants mixed their blood with that of the descendants of Revolutionary War veterans who had settled the country in the 1820s, when the Shawnee Indians were still vigorous. This produced a culture whose dominant verities still were hard work, honesty, church on Sunday and the Republican party.

Neil was the first of three children of Stephen Armstrong, a sometimes bluff but usually gentle and smiling state employee who spent Neil's formative years as an auditor of county records around the state, a job in which he helped send several cheating officials to prison. His mother, Viola Engel Armstrong, was a slim, gracious woman who had always been drawn to music and books, and whose attitudes influenced her son. "My parents are characteristic of the area where I grew up," the astronaut said in Houston. "At the risk of being wrong, it was my observation that the people of that community felt it was important to do a useful job and do it well."

Young Neil began working part-time at age seven, cutting grass in a cemetery in Upper Sandusky, Ohio, for ten cents an hour, and progressing as a teen-ager to stock boy in a Wapakoneta drugstore. This kind of thing was simply expected in Wapak, as the local residents called the town.

"I told all my children I hoped they would pick out something worthwhile, that would do some good for other people, to set their goals high, do the best they could, and they'd have a happy life," said Neil's mother.

Because of the elder Armstrong's roving job, the family moved from one northern Ohio town to another six times during Neil's first six years. But the essential quality of family life remained constant. Mrs. Armstrong put her finger on one of its ingredients: "I was an only child, and I was so thrilled at having the children that I thought I

would never be as near to heaven. Just being around them
was enough for me." Young Neil flourished in an atmosphere of such attention and care.

The hours Mrs. Armstrong spent leafing through maga-
zines with her son, reading books to him and constantly
talking with him produced a very bright little boy who
talked early, read 90 books during the first grade, and
skipped the second grade because he could read on a
fifth-grade level. Later, at Wapakoneta's Blume High
School, Neil flourished in science and mathematics, studying calculus outside of school and teaching science and
math courses temporarily during the illness of Grover
Crites, the teacher who encouraged and guided him in his
advanced studies.

Neil's obsession with flight began on a casual family
excursion to the Cleveland municipal airport when he was
two years old. Four years later, his father took him for his
first plane ride in a Ford Tri-Motor. A year later the boy
built his first ten-cent model plane, the first of hundreds
of models to clutter his bedroom over the next several
years. He worked as a teen-age grease monkey at Wapakoneta's small airport. He paid $9 an hour for flying
lessons, and won his pilot's license before he was licensed
to drive a car. He built a wind tunnel in the basement of
the Armstrongs' white clapboard house. He collected
vintage issues of the magazine *Air Trails,* and was as
upset about their loss as about anything else when, years
after he became an astronaut, his house burned down in
Houston.

Aside from aeronautics, young Neil filled up his time
with Boy Scout activities, reading, playing baritone horn
in the school band, learning the piano and making his
first forays into space science. Several such forays were
made in the backyard observatory of Jacob Zint, Wapakoneta's amateur astronomer. With his telescope, Zint could
bring the moon within 1,000 miles. "Most of the kids
would look for two or three minutes and that would be
enough," Zint recalled. "But Neil would look and look and
look."

Always small, younger than most of the others in his
classes and looking even younger, Neil developed into a
shy, nonassertive, not particularly athletic boy. His younger
brother, Dean, recalled that although Neil moved in a

small circle of close friends who went to the normal teenage parties of the day, he seldom had dates in high school; that he went away to Purdue University in Indiana at age 17 an immature, withdrawn youth; and that he grew physically into a man and developed an underlying self-confidence only when he left Purdue after two years to become a Navy combat pilot.

Neil, the youngest man in his squadron, flew 78 combat missions off the carrier U.S.S. *Essex* during the Korean war, including one in which a cable stretched across a North Korean valley—the one made famous in James Michener's *The Bridges at Toko-Ri*—clipped off the wing of his jet. He won the respect and admiration of his older squadron mates by nursing the plane back over friendly territory, then bailing out safely.

A bigger, stronger, more experienced, but still reserved and youthful-looking Neil Armstrong returned to college after Navy service, then left Purdue in 1955 with an aeronautical engineering degree and joined the National Aeronautics and Space Administration (then the National Advisory Committee on Aeronautics).

Armstrong spent the next seven years at Edwards Air Force Base in California becoming one of the most accomplished test pilots in the world. Characteristically, perhaps, he and his wife, Janet, chose not to live in the nearby town of Lancaster, where most of the test pilots lived; instead they acquired and restored a former forest ranger's cabin in the isolated foothills of the San Gabriel Mountains. Those years were marred by the death of one of the Armstrongs' three children, Karen, of a brain tumor.

If anyone was a natural to become an astronaut, Neil Armstrong was it. He was assigned as a pilot on the now-defunct Dyna-Soar project, in which he was to fly a craft that was to have been part spacecraft and part airplane. Apparently anticipating the end of Dyna-Soar, which finally came in 1963, he applied for the astronaut corps. In 1962 he became the first civilian to be admitted to the corps.

On the job, according to those who worked most closely with the crewmen during their training for the moon landing, Armstrong displayed pride in his skill at the manual art of flying, in his ability as a man to master the machine. His flight aboard Gemini 8 in March, 1966, was a testimony to his skill and courage. When maneuvering thrusters ran wild, causing the spacecraft to tumble out of control, Armstrong took charge and his superb piloting saved the lives of himself and his crewmate, Dave Scott. He was said to be infinitely patient. No one could recall ever seeing him lose his poise.

Off the job, Armstrong was what one associate called a "home and hearth" man, a homebody in the comfortable Houston suburbs near the Manned Spacecraft Center. He liked to listen to a wide variety of music, enjoyed fishing and sailplaning, and sought utter privacy with his wife and two sons. Unlike his crewmates, he had an unlisted telephone number. He and Janet attended parties where Neil often stood back and declined to mix at first, but later warmed up and became the last to leave. He did not smoke. He drank, but was never visibly drunk. He kept his own counsel so effectively that even his parents did not know what his philosophical and religious beliefs were.

Aldrin

Edwin Eugene (Buzz) Aldrin, Jr., who was to join Armstrong on the moon's surface, gave an immediate impression of suave urbanity during an interview at Houston. His disappearing blond hair, once thick and curly, was closely cropped, accentuating protruding ears. But the graying sideburns he sometimes wore countered the stereotyped image of The Astronaut. During the chat, he wore a stylish subdued aqua suit with a crisp handkerchief in the pocket. He flashed two rings on the right hand and one on the left, crossed one knee over the other, and sat toying with a pipe as he talked.

Behind this urbane demeanor was an intelligence that very nearly enabled him, some of his colleagues said, to compute orbital maneuvers in his head. The suave manner covered a fierce ambition and an enthusiasm so intense that Colonel Aldrin had been known at times to irritate associates and friends with his assertiveness and compulsion to talk business. And the stylish clothes hid a marvelously conditioned body whose individual muscles were under disciplined control, an asset which had made Aldrin probably the most accomplished of the six Americans to walk in space.

Aldrin was born of mixed Swedish, Dutch, Scottish and English ancestry in the affluent, serene New York City suburb of Montclair, New Jersey. He lived there for the first 17 years of his life, a relatively sheltered youth in a relatively sophisticated environment.

He grew up as the highly prized only son of a very proud man in a household dominated by six females. His father, Col. Edwin Eugene Aldrin, Sr., U. S. Army retired, had himself accomplished much in aviation and had brushed the fringes of fame. He had learned physics at Clark University under Robert Goddard, the father of modern American rocketry. He had taught himself to fly airplanes, served at intervals as military aide to Gen. Billy Mitchell, whom young Buzz Aldrin met as a boy of five, and had started what became the Air Force Institute of Technology in Dayton, Ohio. In 1929, the year before his son was born, he had set a cross-country flying record of 15 hours and 45 minutes.

Colonel Aldrin's work kept him away from home much of the time. For two years, when Buzz was a budding adolescent, he was gone altogether, having returned to active duty in World War II. Edwin Jr. was surrounded most of the time by two older sisters, who let him get away with nothing; a mother who was at once affectionate and calculating in letting her children always know where she was so that they would be secure, and insistent that her children do their best; a grandmother, whose last name coincidentally was Moon; a woman Negro cook, and a Negro nursemaid.

Mrs. Fay Potter, now of Cincinnati, Ohio, 20 months older than the astronaut and the younger of his two sisters, was responsible for her brother's nickname. As an infant she could not pronounce "brother." "Buzzer," as it came out, stuck, and was shortened to "Buzz" when the boy was about ten. Mrs. Potter remembers Buzz as "a typical rascal boy, a pest." She described him as "forever moving, the kind of kid that would drive a mother crazy, I would think."

"He was very active, but easy to manage," recalled Mrs. Alice Howard of Montclair, who was Buzz's nursemaid for most of his first nine years of life. Despite the

boy's easy manageability, she said, he had a tendency
suddenly to run off, disappear and cause great anxiety.

Buzz Aldrin was kind of hyperactive and athletic from
the start. As a young boy he set up hurdles in the backyard of the three-story home next to a park in Montclair.
He walked on stilts, set up a chinning bar and did calis-
thenics. Much later, Mrs. Potter said, Aldrin did "the
most fantastic exercises" on the beach. When the popular
Royal Canadian Air Force exercise book was published,
an adult Buzz Aldrin turned to the last page, reserved for
champion athletes, and zipped through the whole series.

The boy had his share of traumas. There was the time
he came home and found that the adult white mice he
kept had eaten their young. He burst into tears. There
was also the time a big wave knocked him down on the
beach, causing him to fear water for a long time afterward.
But mostly he was a happy, healthy, blond boy who was
able and willing to respond to his mother's expectations.
"Mother was very demanding of excellence, in a nice way,"
Mrs. Potter said. "She provoked us to do our best, but
she never attacked our self-respect."

When Buzz was pushed into first grade a year early
because he was bored with the second year of kindergarten
then required in Montclair, he "strained every nerve to
prove he could make good," recalled Miss Rita Hogan of
Montclair, the teacher who inherited Buzz in the second
grade. "In school," said Miss Hogan, "he was all business."

The astronaut himself believed the seven summers he
spent at a boy's camp in Maine had much to do with
shaping his personality. "It was the competition, the
athletics, being with people," he said. "Schooling didn't
really provide that."

The Buzz Aldrin of high school days took a full part
in the social activities of the time, but there was much
unusual about him. These aspects were recalled shortly
before the mission by his former football coach, his former
mathematics teacher and a former classmate.

The coach, Clarence Anderson, who asked the 140-
pound Buzz to switch from halfback to fill the vacant
center spot on what turned out to be a state championship
team—"He accepted it as a matter of fact, like 'I'm a
team man and here we go.' He was clean, crisp and quick,

and he never made a bad pass from center. He had back-bone, heart and guts."

William Filas, the math teacher at Montclair High, from which Buzz graduated in the top ten percent of the class—"He seemed to enjoy success where you can see success, as in solving math problems. There was no attempt to out-shine other students on a commonplace problem, but when it came to deep analysis and thought, he had this drive to excel. He was a natural in mathematics, and he gave the appearance of being absolutely sure of himself at all times."

Allen Dumont of Montclair, Buzz's former classmate, who went on to become a hunter of executive talent for a major management consulting firm—"He was the most physically and mentally disciplined guy I've known, then and now. He spent his time on serious things. There was not much clowning around."

These qualities enabled Aldrin to graduate third in the class of 1951 at the U. S. Military Academy. Having grown up in an aviation household and gone to West Point at a time when the Air Force was viewed as the coming service, Buzz decided to become a pilot. Eventually he wound up flying an F–86 Sabrejet in the Korean War, in which he shot down two MIG–15's.

Jack Waite, a North American Rockwell Corporation official in Houston who had known Colonel Aldrin for years, summed up the Aldrin of the Air Force this way: "Gung ho, eager, always trying to get the most out of the plane, always trying to get the highest gunnery score."

Buzz Aldrin won a doctor of science degree in astro-nautics at the Massachusetts Institute of Technology in 1963, and soon thereafter, as an expert in orbital rendez-vous, was assigned as an Air Force representative at the Manned Spacecraft Center. It was only a short step to astronaut status in 1964.

During his Apollo training, Aldrin became known as a computer man who took pleasure in making the machine work for him. One of his best-known feats prior to the moon mission was his record 5½-hour walk in space during the Gemini 12 flight in November, 1966. Aldrin was stubborn in his convictions and prone to tell people frankly what he thought. His constant display of knowl-edge, his unending stream of suggestions for better ways to do things, let him in for much good-natured ribbing—as in his nickname, "Dr. Rendezvous."

According to his sister, Mrs. Potter, Buzz Aldrin was so thoroughly wrapped up in his work that he sometimes found it difficult to give full attention—psychologically—to family matters. He and his wife, Joan, were said to attend parties more frequently than the other Apollo astronauts. On some such occasions, according to Jack Waite, when Colonel Aldrin really relaxed, "he has some drinks and it oils his mouth real good." Mostly, he talked business. His extracurricular activities included being a Boy Scout merit badge counselor and an elder and trustee of the Webster, Texas, Presbyterian Church. His hobbies were running, scuba diving and high-bar exercises.

Collins

During the Houston conversations, Michael Collins, pilot of the Apollo 11 command ship, was as open and breezy as Neil Armstrong was reserved. He was slight, almost delicately built, with merry brown eyes and slightly receding dark brown hair. He was in shirtsleeves during the talks, with a pencil behind his ear. He sat down, threw his leg over a chair arm, and talked animatedly. "Stay casual" had always been his unspoken motto.

This casual manner seemed to mask much. Hidden was the fact that Mike Collins's existence was drifting and unfocused until fairly late in life, leading some of his colleagues to expect that he would never do anything special. But always lurking beneath the casual exterior were a hard competitiveness and determination. These qualities, which made him the untouchable handball champion among astronauts, had also made him an astronaut in the first place, once a late-blooming excitement about space flight gave his life an organizing focus. With his ambition fired, Collins applied his underlying intelligence and determination. "And *voilà*, here I am," he said at Houston. "I don't know where I go from here."

Mike Collins was born in a lush apartment house off the scupture-studded Villa Borghese in Rome, one of Europe's most beautiful parks, of an ancestry extending back to County Cork, Ireland (on his father's side) and into pre-Revolutionary America (on his mother's). The youthful Michael was as insouciant as Buzz Aldrin was intense and disciplined, and as drifting and undirected as Neil Armstrong was single-minded.

Such a personality might at first glance have seemed unlikely in someone from so military a background. Mike Collins happened to be the son of an Army general (the late James L. Collins, who was military attaché in Rome when Michael was born), nephew of another (J. Lawton Collins, former Army chief of staff) and brother of yet another (James L. Collins, Jr., commander of 5th Corps artillery in Darmstadt, Germany).

But the Collins household was lively and cultivated, free of military stiffness. Michael's father was a short, athletic man who spoke his mind when he felt like it, played polo, and was able to do handstands at age 65. He was Gen. John J. Pershing's aide in the Philippines, served in the Mexican campaign against Pancho Villa and with the American Expeditionary Force during World War I, and was present at the coronation of King George VI. From his father, Michael got the athletic bent that made him a quick, aggressive captain of the wrestling team at St. Albans School for Boys in Washington.

From his mother, a slight, soft-spoken lady, he got affinity for books and music. "Mother studied a lot," said one of Michael's two sisters, Mrs. Cordie Weart of Merritt Island, Florida, near Cape Kennedy. "She learned to speak Italian in Rome, and she saw that we spoke the languages, too. She used to take us to museums and churches and places like that."

By the time Michael entered the fashionable St. Albans School at age 12, he had lived in Rome, in Oklahoma, on Governor's Island in New York Harbor, just outside Baltimore on Chesapeake Bay, in Texas and in Puerto Rico. In San Juan, where his father was commander of the Army's Puerto Rico Department, the family lived in a great, sprawling, 16th-century mansion called Casa Blanca, and Michael went to a private school.

"I think a life like that is good for a kid," astronaut Collins said of the constant moving about. "You learn some things twice in school, and others not at all, and you have to leave a lot of friends. But on the other hand, it's a damned interesting life for a kid. On balance, I think it's an advantage."

The family's gypsy life caused its members to grow closer in support of each other, Mrs. Weart believes. Her mother added, "We did a great many things together. We were quite a close family, we really were, and I thoroughly enjoyed my children."

Because he was by far the youngest of the family—his brother, for example, was 13 years older—Michael was its pet. He was also subject to a subtle love-laced discipline that, in the astronaut's words, caused him to "go to great lengths to avoid doing anything to displease my parents. To gain their approval was important to me."

But the parents were not pushers, as was evidenced by Mrs. Collins's attitude on the children's careers: "We just told them they were to do what they wanted. Neither one of us thought we should live our children's lives."

In such an atmosphere, young Mike did what was expected of him, did not attract particular attention from those looking for achievement, and enjoyed the swimming, fishing and playing that make up a normal boy's life. Carrying his low-voltage demeanor to St. Albans, Mike Collins quickly became one of the most popular boys there, one of the least dedicated and one of the most mischievous. "I've got to say I just didn't like school," he explained later. "I think St. Albans let me in sort of to be nice."

Boarding students at St. Albans lived an almost Spartan existence in small cubicles that resembled monks' cells. But within this atmosphere, Mike Collins was known among the masters as a perpetrator of michief. To this day, they said, they were unable to pinpoint exactly what he did, but they were always sure he was the spark for much of what went on behind the scenes. "He had an immobile face," said Ferdinand Ruge, one of the masters. "You always wondered what he was concocting."

Mike served as a prefect—one of the top student leaders—during his senior year, and the St. Albans yearbook for 1948 noted that he was one of the four most popular boys in the school. It noted also that a little knowledge might have seeped into him "despite the lures of Morpheus (the Greek god of sleep)." His sleepiness may have been caused by the fact that he had to get up to be a Cathedral altar boy at the 6:30 A.M. service.

His grades at St. Albans were not spectacular, although

he made the equivalent of B-pluses in mathematics, his best subject. John Davis, now assistant headmaster and a former teacher of Mike's, noted "an intellectual precision" and "a desire to see the limits of a problem—then he got to the heart of it."

Michael Collins graduated from West Point in 1952 with a less than brilliant academic record. The yearbook noted that his battle cry was "Stay casual," and that "he took the cash and let the credit go and seldom heeded the distant rumble." Freely translated, that meant "live today and don't worry about tomorrow."

What led Mike Collins into the Air Force is not exactly clear. But once in, the astronaut said, he was interested in flying the newest kinds of planes. So he became a test pilot. "To avoid being static—that's why you do something like that," he explained.

"Even after he came here [to Edwards Air Force Base] to test-pilot school, I thought he was just going to be another guy who was going to drift through life," said Bill Dana, a NASA test pilot who knew Collins well at West Point and was his roommate in flight training. "I think the space briefings here really turned him on, and he bloomed."

Collins's Apollo colleagues described him as being "very understanding—perhaps too understanding. He was a nice fellow who did not want to make people feel bad. Regarding his big chance on Apollo 11, Collins was said to be fascinated by the wide variety of tasks required of the Command Module pilot, who must run an entire spaceship by himself. His preparation for the moon flight included Gemini 10, during which he walked twice in space and took part in a rendezvous with two separate space vehicles.

Of the three Apollo 11 men, Collins was perhaps the most stay-at-home. He was said to be quick to admire people, impossible to dislike, so considerate that he would not say anything unpleasant to anyone. He drank martinis, sometimes several of them, but with no apparent effects. He had a passion for fishing; and he declined to fertilize his lawn so that he would not have to cut it. Sometimes he could be seen hopping around trying to catch the family's white rabbit as it munched clover. He and his wife, Pat, and their three children formed what a neighbor said was "a very close family circle."

Middle-class Supertechnicians

All three Apollo 11 astronauts—Armstrong, Aldrin and Collins—were born as the nation skidded to the bottom of the Great Depression in 1930: Aldrin on January 20, Armstrong on August 5, and Collins on October 31. But all three were sheltered from the ravages of that depression. None had ever come close to privation. All three were expressions of the dominant values of the broad American middle class, but each represented a different current in that mainstream of society.

All three—because of their time in history and the skills developed out of their middle-class heritage—also were among the supertechnicians of their day. They were not scientists seeking fundamental truth—although astronauts on future flights would be—but supremely self-confident pilots who liked action; and highly disciplined engineers whose natural habitat was the sometimes bewildering technology of the electronic age.

From one point of view, they were the most flexible and versatile components in what was perhaps the most complex technological system ever devised. But they were also in many ways ordinary men, who, despite an extraordinary psychological stability, displayed very human foibles of personality.

It was not surprising, therefore, that the men of Apollo 11 were beginning months before the mission to be concerned about the consequences of the fame that surely awaited them if they carried out the moon landing and return to earth. Mrs. Potter said that her brother, Colonel Aldrin, was starting to worry about loss of privacy and freedom. Colonel Collins's mother said her son was slightly irritated that because of him, his family should be subjected to public scrutiny and prying. "He doesn't think an old lady like me should have to put up with that," said the 73-year-old Mrs. Collins.

The astronauts' concern about such things was one of the latest facts of existence for all three—and one thing was certain: If they succeeded in the epic mission on which they were to embark, all that was in their personal past would be merely prologue, and life for them—and perhaps the world—would never be the same.

Chapter XVIII

Days of Waiting

Throughout the history of exploration there had never been an expedition planned and prepared with such care as Apollo 11.

The many and highly complex components of the Command Module, the Lunar Module and the Saturn 5 had been tested in a variety of ways, both on the ground and in space. For years men and computers had weighed the equations, plotted the course, anticipated the risks and analyzed ways to minimize them. Their work had led to ways to keep the spacecraft from getting too hot or too cold, keep it from losing communications with earth, keep its engines from firing too long or too short, keep its course at distances of 250,000 miles from its home base.

Every pound of weight in the rocket and spacecraft— the moon vehicle that was standing on Launching Pad 39-A at Cape Kennedy in July, 1969—had met the test of usefulness. John P. Mayer, chief of mission planning and analysis at the Manned Spacecraft Center in Houston, once described how carefully this was analyzed. To determine that the astronauts had enough propellant to fly the mission but were not carrying excess weight, it was necessary to develop mathematical models of the sun-earth-moon environment and of all the spacecraft systems. Using these models, Mr. Mayer and his associates flew hundreds of imaginary missions in the computers.

"The computer programs used to fly these imaginary missions are called the Monte Carlo programs," Mr. Mayer explained. "They contain thousands of possible Apollo 11 missions, from the worst possible missions to the best possible missions. The computer program then picks several score of these missions at random, and in this way the probability of the actual mission's success can be determined.

All these problems must be solved and a nominal course to the moon—the one flown if everything works perfectly—must be found."

Dress Rehearsal

Such careful preparation continued right through those final days of waiting, the days leading up to launching time on July 16.

Early on June 27, launching crews began the Countdown Demonstration Test (CDDT), or the dress rehearsal, for Apollo 11. This included all the steps of the actual countdown, all the tests, the complete fueling of the rocket, the entrance of the astronauts into the spacecraft cabin. A week later, shortly before 9:32 A.M. on July 3, a clear voice came through the intercom at Firing Room 1: "You're cleared for firing command." In a few seconds, a man wearing a white shortsleeve shirt leaned over his control console and pressed the button that initiated the automatic, three-minute and seven-second sequence of computerized events leading up to lift-off. Without a hitch, the countdown ticked away to the point of ignition. But it was only a make-believe countdown, the final practice for the real thing.

"The countdown test is my green light, once it is satisfactorily completed," Rocco A. Petrone, the launch director, said. "Only at that time do I know we have accomplished all our checkout procedures and that I am ready to give a go for the real countdown."

The three astronauts—Armstrong, Aldrin and Collins—were also busy during those days of waiting. Early one morning, less than a week before lift-off, I went to the flight crew training building. From a glassed-in observation area, I watched the three men dressed in slacks and cotton sports shirts slip out of their loafers and climb the steps to an odd-looking contraption known as the Apollo Mission Simulator.

Outside, the 40-ton, 30-foot-high structure of brown metal bulged at strange angles to cover the television cameras, five tons of mirrors and lenses and racks of complex electronic equipment—all of which was designed to duplicate the sounds and sights, maneuvers and malfunctions of a moon mission. Inside, the cockpit was an almost exact replica of the Command Module. All the dials and gauges, switches, hand controls and computer display panels were just as they are in the real spacecraft.

For nearly eight hours, while instructors operated the four computers that ran the simulator, the astronauts practiced over and over again the first 12 minutes of the flight. There were times when the make-believe lift-off was free of trouble. Other times, to test the astronauts' reflexes and knowledge of the systems and alternate mission plans, the instructors ordered the computer to throw in a malfunction or two. The computers were capable of introducing 1,700 malfunctions for the astronauts and the flight controllers, who were involved in the practice session at their consoles in Houston, to cope with. The instructor's job, Riley D. McCafferty, the director of simulations, told me, is "to study the men, see where they are weak and where they are strong and to see if the procedures for the mission are adequate."

For the last two months before the flight, each of the astronauts spent more than 400 hours in Command Module and Lunar Module simulators at Cape Kennedy and at Houston. It was thus possible to practice every step of the eight-day mission—launching, earth-orbit operations, the rocket firing to leave low earth orbit and head for the moon, the insertion into lunar orbit, the landing on the moon, the lift-off from the moon and the return to earth.

One of the most unusual simulators was a real flying vehicle, the lunar landing training vehicle. It had a Rube Goldberg look about it and a nickname to match—"the Flying Bedstead." But it had much the same pilot-handling characteristics as the Lunar Module; it even had a turbojet to offset five-sixths of the vehicle's weight and thus approximate the one-sixth gravity of the moon. Undisturbed by an accident earlier in the year, in which Armstrong had to bail out of a crashing trainer, the two moon explorers spent hours piloting the vehicle to soft touchdowns.

All three astronauts also devoted long hours to learning their mission plan, operating the guidance and navigation computer (which took nearly 40 percent of their training time), experiencing a lunar vacuum in the large vacuum chamber at Houston and getting brief sensations of weightlessness during sudden dips in the KC–135 airplane. But in the last few days before lift-off, the three men tapered off their training. They confined themselves to reviewing flight instructions and taking a few runs in the computerized simulators.

The Disinvited President

They were following their doctor's orders. Having in mind the fatigue and illness that delayed Apollo 9 by three days, Charles A. Berry, the astronauts' flight surgeon, ordered the training slowdown and also placed the crew in semi-quarantine for the final two weeks. The men were watched for the slightest symptom of illness that might hamper the flight. The only people who could come in contact with them were their training associates and their immediate families—so long as they were healthy.

No exception could be made even for the President of the United States. After the arrangements were already made for President Nixon to dine with the astronauts on the eve of their launching, Berry "suggested" to the White House that, from a medical standpoint, the President and the germs he might be carrying were not welcome. The dinner was cancelled.

Another kind of pathogen—a disease-causing agent—was the object of far more elaborate and restrictive quarantine measures being prepared for the post-flight period. At an $8.1-million Lunar Receiving Laboratory in Houston, scientists were waiting to receive the returning astronauts like the plague. This laboratory contains the sealed quarters to which the astronauts would be taken in a sealed van after their splashdown. Such isolation was ordered on the off-chance that the astronauts would bring back from the moon some organisms against which man has developed no immunity. The result could conceivably be a catastrophic plague on earth.

Just as the astronauts were waiting for their moon journey, the possibility of "back-contamination" was dramatized in a chilling best-seller, *The Andromeda Strain*. The book was a fictional account of the consequences of a strain of alien germs brought back to earth from out in space.

The Apollo 11 astronauts were not likely, however, to bring back some deadly strain. Biologists and epidemiologists doubted that the airless moon harbored any such live organisms. Nonetheless, early in the Apollo project, a committee of the National Academy of Sciences had decided that a strict quarantine was desirable—just in case. This led to NASA's construction of the laboratory with its vacuum chambers for the lunar dirt samples and hermetically sealed quarters for the astronauts and their doctors, technicians and cook.

Whatever organisms the astronauts brought back from the moon, they were not starting their voyage with any detectable signs of illness. On Friday, July 11, they underwent a four-hour medical examination and were pronounced physically fit for flight.

Countdown

No magic button was pressed, or switch flipped. A master clock in Firing Room 1 simply began ticking at 8 P.M. on Thursday, July 10. As the clock ticked, launching controllers began routing electrical power to the 363-foot-tall moon vehicle at the pad. It was the beginning of the countdown. The days of waiting were now being measured in hours and minutes and seconds.

While the astronauts remained in seclusion and the launching crews ministered to the rocket and spacecraft, the crowds streamed into Cocoa Beach and the other communities surrounding Cape Kennedy and the moonport. They were the 5,000 very important persons—diplomats, congressmen, judges, mayors, bankers, industrialists and other assorted celebrities—invited by NASA. They were the 3,000 journalists, many of them foreign. They were the Australian marathon runner who jogged all the way from Houston and the Rev. Ralph D. Abernathy with his demonstrators calling attention to the needs of the poor people of the country, needs that many felt were slighted partly because of the expense of the moon project. The crowds also included the executives of the Apollo contractors, their public relations men, school groups and ordinary tourists who wanted to tell their children's children that they had seen the first men go to the moon.

Bernard Weinraub described the gathering of the crowds and the building tensions in a story for the *Times:*

This coastal city—spawned in the space age—waits with jittery excitement.

"Hell, this is it," exclaimed Charles R. Johnson, the manager of the Convention and Visitors Bureau for the Cape Kennedy Chamber of Commerce. "After 14 years here, it's almost as if I were a kid again looking forward to Christmas."

Nearly one million tourists are expected to pack this central Florida community for Wednesday's Apollo 11 moon launching—an influx that is expected to provide $4- to $5-million to financially worried Brevard County.

Everywhere—in the stores crammed with $2.98 Apollo 11 toys, in the restaurants that offer $1.25 "lift-off" martinis, at supermarkets with signs reading "We will open all night before the launch"—this town of 13,500 and surrounding communities are riveted for Apollo 11.

"The excitement doesn't even have to be talked about —it's in the air," said Dr. Burton Podnos, a 38-year-old psychiatrist, who is director of the county's mental health center. "The kids are acting out the excitement and tensions of parents who are involved in the space program. We have more calls from parents now saying, 'I don't know what to do with Johnny, he's uncontrollable this week.' "

Virtually every day now, policemen, civil defense and space agency officials hold meetings and confer by telephone to make last-minute changes in the plans for launching day—plans on how to deploy the 1,000 policemen and volunteers, the 30 ambulances, the army helicopter, even the 10 city boats and the Coast Guard squadron that will prowl the Banana River where 1,000 boats (with such hard-earned names as "Miss Overtime" and "Miss Extra Hours") will line up to watch the lift-off.

Officials are bluntly nervous about the 300,000 automobiles expected to surge into the county. "That's the equivalent of 1,000 miles of automobiles parked bumper to bumper here," said Herbert W. Johnson, the director of the county civil defense headquarters, whose below-ground facilities are the traffic and safety command post. "Well, we only have 1,000 miles of highways here, so physically, cars could cover every road in the county."

On this hot, breezeless day in Cocoa Beach, the growing influx of visitors was affecting virtually every mood and decision, every conversation, every age group, every resident in the city of sand dunes and palmettos.

Near the Galaxy Lounge of George's Steak House— replete with phone booths shaped like Gemini capsules and bar stools that have seat belts—Walt Crosley, a part owner, was ordering 3,000 steaks over the phone and extra supplies of bread and milk for early next week.

"We feel there won't be enough food in the area," said the 31-year-old restaurateur, echoing the feeling of other businessmen. "Food trucks won't be able to get in and we've got to store up now."

The 30 motels in Brevard County have been sold out —some for nearly a year. Motels as far away as Daytona, 75 miles north, and Orlando, 60 miles west, are also packed. Within Cocoa Beach, most of the 300 homes that have offered to take guests—some free, some for $20 or $25 a night—are booked.

"We're getting phone calls from all over—Europe, South America—all over," said Mrs. Maybell Wilkins, a gray-haired, Missouri-born employee at the Chamber of Commerce, placing a phone on the receiver and listening to it ring 30 seconds later.

"Just yesterday I called the Palm East Apartments and I told the girl there that the president of the Bank of Nevada needed a room and she just laughed and said they couldn't even give Darryl Zanuck a room," said Mrs. Wilkins. "Can you imagine?"

The shortage of hotel rooms outside Cocoa Beach even had an impact on hundreds of Europeans who hoped to take a "moon shot tour" that was sponsored by Trans World Airlines, five European newspapers and the Taylor Travel Service of Paris.

Although 375 French, German, Spanish, Swiss and British tourists are scheduled to arrive in the United States on Monday for the tour that includes a view of the launching, hundreds more were turned away because of a shortage of hotel space in Orlando, where the visitors will spend Tuesday night. "The response was just overwhelming," said a T.W.A. spokesman.

Virtually every type of business is now flourishing in Brevard County as a result of the moon shot. Motels have purchased and rented cots and deck chairs to line swimming pools on Monday and Tuesday nights for visitors unable to find rooms.

Such bars as Ramon's, the Mouse Trap and the Missile Lounge, whose T.G.I.F. (Thank God It's Friday) parties are a Cocoa Beach tradition, now advertise "T.G.I.F. parties every night," where combos play, of course, "Fly Me to the Moon."

Even out-of-state surfboarders—who usually swarm onto Cocoa Beach in the autumn when hurricane waves lash the surf—have driven and hitchhiked here.

"We've rented and sold everything out and it's not even the season," said 16-year-old Ron Rubino, a proprietor of "Soul Surfboards" on North Atlantic Avenue. "Most of the people renting them are older, guys who are 25, even 30."

Memories

Two men who arrived at Cape Kennedy for the launching evoked, by their presence, memories of the time in 1961 when the moon project was initiated.

As senator, vice-president and president, Lyndon B. Johnson was one of the most ardent supporters of a strong national space effort. He sponsored the bill that created NASA. And in April, 1961, he presided at a series of high-level government meetings out of which came the decision to go to the moon. Now he was at the Cape, at the special invitation of his successor, to see the fruition of much of his work.

Another guest was James E. Webb, the administrator of NASA from 1961 through the Apollo mobilization until 1968.

If they were reminders of Apollo's earlier days, another event occurred during the countdown to bring back memories of the atmosphere in which Apollo came into being. The Soviet Union, the nation that pioneered in space and for many years made it a habit to be first on that frontier, launched its own moon ship. Although it was unmanned, Luna 15 sent a shudder through Cape Kennedy. What if it were an effort to upstage Apollo 11? What if it were the long-rumored automated vehicle designed to land on the moon, scoop up some dirt and bring it back to earth? No one knew, and the Russians remained silent. But if it were a moon scooper, Luna 15 might just be able to give the Russians the glory of being the first to return a bit of another world.

At any rate, it kept Americans guessing as they waited for Apollo 11.

"Willing and Ready"

The countdown proceeded smoothly. Petrone called it the smoothest yet. The crowds grew larger. It was hard to believe but it was happening: men were poised to go to the moon.

Two nights before the launching, Armstrong, Aldrin and Collins made their final public appearances in a nationally televised 30-minute interview. They were asked questions by four newsmen selected from the 3,000 there. Because of the semi-quarantine, the interview was conducted over closed-circuit television. The astronauts, wearing short-sleeved shirts and sitting in soft chairs, were in one building and the newsmen were 15 miles away in another.

Armstrong appeared slightly nervous, choosing his words with laborious care, and Aldrin was stiff and usually unsmiling. Only Collins appeared relaxed.

Asked if they had any fears about the mission, Armstrong replied: "Fear is not an unknown emotion to us. But we have no fear of launching out on this expedition."

Aldrin told the newsmen that it was all right to use the term "when" and not "if" in referring to the moon landing. "We're thinking positively," he said.

"After a decade of planning and hard work," Armstrong declared, "we're willing and ready to attempt to achieve our national goal."

Chapter XIX

Four Days to the Moon

"Go, baby, go!" The familiar murmured exhortation spread like fire through the control rooms. Only, this time they said it with more fervor than usual. On the beaches they shaded their eyes against the bright sun and held their breaths. Across the land they paused at their jobs and rushed to television sets.

This was the "big one" at last. It was not just another space flight. This was *the* historic lunar odyssey to put men on the moon! It was the climax of a decade of work and a score of preparatory missions. Could Armstrong and Aldrin really land and walk on the moon? The whole world waited anxiously to find out.

It was July 16, 1969, and the "big one" began on schedule. At 9:32 A.M., E.D.T., orange flames and dark smoke spewed out of the huge Saturn 5 rocket supporting the Apollo 11 spaceship at Cape Kennedy. Then, ever so slowly, the 3,817-ton vehicle struggled to overcome the earth's gravity, finally cleared the Pad 39-A launching tower, and arced out over the Atlantic Ocean.

As the vehicle flawlessly propelled Neil Armstrong, Buzz Aldrin and Mike Collins into their eight-day, 500,000-mile journey, the powerful blast-off sent a tremor through the ground, and staccato shock waves beating at the estimated one million people—a record for a launching—who stood under the Florida sun that Wednesday morning to see the start of the most daring mission thus far in the age of spacefaring. Millions more, possibly half a billion, across the nation and the world watched over television relayed by satellites.

"C'est formidable," a young French tourist at the Cape whispered as Apollo 11 streaked on a tail of fire through a wispy white cloud. "C'est magnifique."

In Firing Room 1, three miles from the pad, a proud cheer went up from the 400 men who had directed the countdown. In this spacious rectangular room, jammed with consoles and gauges and a master computer, were the specialists who manned all the technical systems that had to mesh with almost incredible precision for the mission to be "go." Some knew every intricacy of the engines; others were responsible for guidance, for stabilization equipment, for computers, for communications.

It was the general view that the operation in Firing Room 1 on July 16 was the most professional so far and, at the same time, the tensest. "I'd have to say," asserted Rocco A. Petrone, the launching director, "that you would feel the tension. The people knew this was the big one. There was a certain amount of, let's say, static electricity in the air."

Space-Borne

For three minutes the blast-off cast a spell over the spectators on the Cocoa Beach sands, in the V.I.P. stands, in the armada of small boats clogging the Banana River. It was like a microcosm of America—the powerful and the powerless, the scientists and salesmen, the vacationing families, the journalists, the quick-buck hawkers, the scholars, the young and old—all thrown together by a common interest and anxiety. Some applauded; others simply stared.

Then, as the flight proceeded through the critical first steps without a hitch, relaxation and mixed reaction set in. Jack Benny cracked jokes at the V.I.P. site. A university student exclaimed, "This is something I'll tell my grandchildren about—I was here when it happened!" Lyndon Johnson, who had pushed the moon program during his years in the Senate and the White House, wondered "why we can't put the same effort into peace for all time." A group demonstrated for the poor as a small reminder of the social problems down here on earth. But perhaps the dominant reaction was expressed by Dr. Kurt H. Debus, director of the Kennedy Space Center, who said, "We have left the land and taken over the air and have moved mankind really together."

President Nixon, who had watched the lift-off on television in Washington, declared a National Day of Partici-

pation for the lunar exploration set for four days hence.
Urging that Americans be given the day off from work,
he declared:

"In past ages, exploration was a lonely enterprise. But
today, the miracles of space travel are matched by miracles
of space communications; even across the vast lunar distance, television brings the moment of discovery into our
homes and makes all of us participants. As the astronauts
go where man has never gone, as they attempt what man
has never tried, we on earth will want, as one people, to be
with them in spirit, to share the glory and the wonder and
to support them with prayers that all will go well."

The proclamation was in line with the government's efforts to give all Americans, indeed all mankind, a sense of
being a part of this history-in-the-making. The astronauts
already were carrying with them tokens to be left on the
moon as symbols of the epic achievement. In addition to
the metal plaque attached to a leg of the Lunar Module,
the tokens included messages of goodwill from 73 heads of
state, a United States flag and five medals honoring the
Soviet cosmonauts and American astronauts who have died
in the conquest of space.

It was this feeling of history being made while the whole
world watched, of mankind reaching beyond his own planet,
that accounted for the high interest among a citizenry which,
because of previous space successes, had grown somewhat
blasé about launchings. Apollo 11 was different from other
major feats of exploration. In the past, only a few people,
as a rule, were aware at the time of the great events that
altered the course of man's destiny. Those who set forth
across the unknown seas and continents did so on their own,
and the world had to wait months, often years, for word
of success or failure. Not so with Apollo 11.

Actually, for the first four days of the Apollo journey,
the world saw little that was new. Moving toward the
vicinity of the moon, the three astronauts closely followed
a trail already blazed by Apollo 10 two months before.
Until the landing craft with Armstrong and Aldrin aboard
swooped to within 50,000 feet of the moon where the final
descent began, the crew faced challenges and risks that, in
a sense, were relatively familiar. They encountered no
trouble.

During this period, the astronauts performed the routine

assignments of a space flight, checking equipment and systems, carrying out everyday housekeeping chores, and conducting regularly scheduled telecasts for the folks back home. From all indications, the men were comfortable and relaxed. "We have a happy home," Collins said during one of the telecasts. "Plenty of room for the three of us. We're all finding our favorite little corners."

On Friday, when Apollo 11 was three-quarters of the way to the moon, Armstrong and Aldrin took a televised inspection tour of the attached Lunar Module. To do this, they had to open the hatch at the apex of the cone-shaped Command Ship, named Columbia, and then squeeze through the 30-inch-wide tunnel into the LM, named the Eagle. This was the procedure they were to follow later for the actual descent. The telecast, which ran an hour and 36 minutes, was one of the clearest color TV transmissions ever sent from space. Viewers on earth could read dials on the LM control panels and instruction labels at the point where the two ships were joined.

It was on Friday also that the Soviet Union gave assurances that its unmanned lunar-orbiting spaceship, Luna 15, would not interfere with Apollo 11. The assurances were made in a cable from Dr. Mstislav V. Keldysh, president of the Soviet Academy of Sciences, to Colonel Frank Borman, the Apollo 8 commander who had recently returned from a visit to Russia and had been asked to telephone the Russians for such information. Under the United Nations Outer Space Treaty of 1967, which was established in part to avoid space mishaps, the United States and the Soviet Union are bound to furnish this type of data. This was the first time, as far as American space agency officials could recall, that the Soviet Union had communicated directly with the Americans about one of their missions while it was in flight.

Into Lunar Orbit

The first really anxious moments of the Apollo mission came around noon on Saturday when the linked Columbia and Eagle swung around the leading edge of the moon to begin the transition into lunar orbit. For 33 minutes, the spacecraft was out of radio contact with tracking stations on earth. During that time, the astronauts had to fire the craft's main rocket to slow the vehicle down so that it could be captured by lunar gravity. If the rocket failed to fire, Apollo 11 would loop the moon and head back to earth. If it fired too long, the vehicle could crash into the lunar surface. But the rocket performed almost flawlessly. When the space ship swept into moon orbit, Armstrong broke the tense silence at last: "It was like perfect," he reported.

But everything up to now was really only warm-up for the highly complex descent of Eagle to the lunar surface—the encounter with the unknown. Here was a challenge which men had never tackled before. During a span of little more than 10 minutes, Armstrong and Aldrin had to sweep about 300 miles across the face of the moon in their LM, descending on a long curve from 50,000 feet to touchdown. A major equipment failure or a miscalculation could mean either abandonment of the effort short of the goal or, in the final phase, disaster. If one of the LM's spindly legs broke upon contact because of boulders or an unexpected movement, or the touchdown occurred on a relatively steep slope, the craft could tip over and the astronauts would be unable to lift off for a return trip.

The crucial descent procedure began on the Apollo's thirteenth moon orbit after Armstrong and Aldrin had crawled into the LM, leaving Collins to fly the Command Module. The two ships broke their link and drew apart. "How does it look?" Mission Control asked Armstrong. "Eagle has wings," he replied.

Armstrong and Aldrin rode the Eagle back around to the moon's far side where the descent engine was to fire in order to send the module toward the moon on a long, curving trajectory. Suspense built up in the control room at Houston because once again the astronauts were out of radio touch with the ground. Among those waiting in the room for word of the rocket firing were Tom Paine, NASA's administrator, most of the Apollo project officials and several astronauts.

Armstrong finally broke the suspense with a calm report: "The burn was on time." In the Command Module, Collins reported to Mission Control, "Listen baby, things are going just swimmingly, just beautiful." Then Eugene F. Kranz, the flight director in Houston, turned to his associates and said: "We're off to a good start. Play it cool."

When Armstrong and Aldrin reached 50,000 feet from the moon's surface, green lights on the computer display keyboard blinked the number 99. This signaled Armstrong that he had five seconds to decide whether to go ahead for the landing or return to the Command Module. The ship commander pressed the "proceed" button.

The throttleable engine built up thrust gradually, firing continuously as the LM descended along the steadily steepening trajectory. Seven minutes after the firing, the astronauts, riding upside down according to plan, were 21,000 feet above the surface and moving toward the landing site. The guidance computer was driving the rocket engine. At 7,200 feet, with the landing site about five miles ahead, the computer ordered maneuvers to tilt the bug-shaped craft almost upright. Armstrong and Aldrin then got their first close-up view of the plain they were aiming for.

The panorama rushed below them—the myriad craters, hills and ridges, deep cracks and ancient rubble of the moon. Their target was in the Sea of Tranquillity, one of five smooth landing sites that had been selected along the lunar equator on the basis of pictures taken by unmanned spacecraft. Being on the equator reduces the maneuvering required for the astronauts to get there. The sites were on the side of the moon always facing earth; this, of course, made it possible to communicate with the explorers.

The Eagle closed in on the target, dropping about 20 feet a second until it was hovering almost directly over the selected landing spot at an altitude of 300 feet. Then came a startling discovery: the floor was littered with boulders.

Armstrong quickly grabbed semi-manual control. As he said later, "The auto-targeting was taking us right into a football-sized crater, with a large number of boulders and rocks."

Armstrong retained semimanual control of the module for the rest of the way down. The computer maintained control of the rocket firing, but the astronaut could adjust the craft's hovering position. For about 90 seconds, Armstrong searched for a clear spot.

"Down two and a half," came the radio report, probably from Aldrin. "Forward, forward 40 feet. Down two and a half. Picking up some dust. Thirty feet. Two and a half down. Shadow. Four forward, four forward, drifting to the right a little."

The fuel in the descent engine was now running low. Armstrong had only a few seconds left to make his landing or to abort and return, by firing the ascent engine, to the Command Module. Armstrong remained cool.

Finally, Armstrong found a spot he liked. He eased the Lunar Module down slowly until a blue light on the cockpit flashed to indicate that the five-foot-long probes, like curb feelers, on three of the legs had touched the surface.

He pressed a button marked "Stop" and reported, "Okay, engine stop."

Charles M. Duke, the capsule communicator at Mission Control, radioed, "We copy you down, Eagle."

Then the historic words: "Houston, Tranquillity Base here. The Eagle has landed."

It was 4:17 P.M., E.D.T. The module was at an angle of only about four and a half degrees. The angle could have been more than 30 degrees without tipping the vehicle over.

"Roger, Tranquillity," ground control replied. "We copy you on the ground. You got a bunch of guys about to turn blue. We are breathing again. Thanks a lot."

Although Armstrong is known as a man of few words, his heartbeats told of his excitement upon leading man's first landing on the moon. At the time of the descent rocket ignition, his heartbeat rate registered 110 a minute—77 is normal for him—and it shot up to 156 at touchdown.

The landing actually was 4 miles down range—meaning 4 miles too far to the west—but it was well within the designated area. Several small factors contributed to errors in some data fed into the craft's guidance computer. This is what necessitated Armstrong's taking over manually. When they landed, the astronauts were not sure exactly where they were.

Mission Control immediately reported the landing to Collins who was riding the command ship Columbia about 65 miles overhead. "Yea, I heard the whole thing," replied the man who went so far but not all the way. "Fantastic."

Yes, fantastic! They had done it! By flying their craft to the lunar surface and landing there, by conquering this great unknown, Armstrong and Aldrin had proved that the moon, long the symbol of the inaccessible, was now within man's reach. The next step, the true purpose of this fantastic mission, was to sample the unknown, to walk on the moon and learn how hospitable or hostile it was.

Chapter XX

On the Moon

At 10:56 P.M., E.D.T., on July 20, 1969, Neil A. Armstrong stepped into history. From the bottom rung of the ladder leading down from Apollo 11's landing craft, he reached out his booted left foot and planted the first human footprint on the moon.

Then he uttered the long-awaited words that are sure to be immortalized: "That's one small step for man, one giant leap for mankind."

There it was, man meeting moon, his first direct contact with another celestial body. For explorers, it was the realization of centuries of dreams. For scientists, it meant an unprecedented opportunity for possible clues to the origin and nature of both the moon and the earth.

Appropriately, Armstrong was able to share the triumphal moment with mankind. As he descended the ladder, he pulled a lanyard that released a fold-down equipment compartment that deployed a television camera. Thus, through the miracle of modern communications, hundreds of millions of people on earth—probably the largest audience ever—witnessed the astronaut's memorable step via TV and heard his words via radio. It required just 1.3 seconds, the time it takes for radio waves to travel the 238,000 miles between moon and earth, for Armstrong's image to appear on home screens. This gave viewers a feeling of "I was there" when history was made.

What was this new environment like, this remote space frontier suddenly invaded by man? Looking through the windows of the landing craft, the astronauts saw a bleak but strangely beautiful world. It was just before dawn over the Sea of Tranquillity, with the sun low over the eastern horizon behind them. The chill of the long lunar night still clung to the boulders, craters and hills before them.

"Magnificent desolation," was the phrase Aldrin used in describing the view. He said that he could see "literally thousands of small craters." But most of all he was impressed initially by the "variety of shapes, angularities, granularities" of the rocks and soil around Tranquillity Base.

At one point, Buzz Aldrin radioed this impression of the general area in which they touched down:

> [There is a] level plain cratered with a fairly large number of craters of the 5- to 50-foot variety. And some ridges, small, 20 to 30 feet high, I would guess. And literally thousands of little one- and two-foot craters around the area. We see some angular blocks out several hundred feet in front of us that are probably two feet in size and have angular edges. There is a hill in view just about on the ground track ahead of us. Difficult to estimate, but might be half a mile or a mile. . . .
>
> I'd say the color of the local surface is very comparable to that we observed from orbit at this sun angle —about 10 degrees sun angle or that nature. It's pretty much without color. It's gray and it's very white as you look into the zero phase line. And it's considerably darker gray, more like an ashen gray, as you look out 90 degrees to the sun. Some of the surface rocks in close here that have been fractured or disturbed by the rocket engine plume are coated with this light gray on the outside. But where they've been broken, they display a dark, very dark, gray interior and it looks like it could be country basalt.

When Armstrong reached the bottom of the Lunar Module's ladder, he found that the moon was indeed not made of green cheese. Observing that "the LM foot pads are only depressed in the surface about one or two inches," he said:

"The surface is fine and powdery. I can pick it up loosely with my toe. It does adhere in fine layers like powdered charcoal to the sole and sides of my boots. I only go in a small fraction of an inch, maybe an eighth of an inch. But I can see the footprints of my boots and the treads in the fine sandy particles."

Then, while the excited audience watched those first few moments in awe, Armstrong tentatively tested the moon's environment and found it relatively receptive. He found that he could move about easily in his bulky white spacesuit and heavy backpack while under the influence of lunar gravity, which makes everything weigh only one-sixth of what it weighs on earth.

After 19 minutes, Armstrong was joined outside the landing craft by Aldrin, who had been preparing and handing down equipment for the two hours of probing and experimenting.

The excitement of the moment notwithstanding, Aldrin did not overlook the little necessities. As he emerged through the hatch and started down the ladder, he said, "I want to back up and partially close the hatch, making sure not to lock it on my way out."

"Good thought," Armstrong agreed.

"That's our home for the next couple hours," Aldrin added. "We want to take good care of it."

Then, as Aldrin started his first testing of the surface, Armstrong commented at one point: "Isn't this fun?"

"Right in this area I don't think there's much fine powder," Aldrin noted. "It's hard to tell whether it's a clod or a rock."

"You can pick it up," Armstrong pointed out.

"And it bounces," was Aldrin's reply.

They immediately set up another TV camera away from the craft to give people on earth a broader look at the Sea of Tranquillity landscape. What was seen during a panoramic camera sweep conformed pretty much with photographs previously transmitted by unmanned satellites: a bleak, empty, almost flat, crater-pocked, undulating surface devoid, of course, of vegetation. Yet, Armstrong described the landscape as having "a stark beauty all its own."

"It's like much of the high desert of the United States," he said. "It's different but it's very pretty out here."

One of the first things the astronauts did to embellish that forbidding and monotonous landscape was to plant their 3- by 5-foot American flag. It was stiffened with thin wire so as to appear to be flying on the windless lunar surface.

The moon spectacular began earlier than originally scheduled, but a bit later than Armstrong and Aldrin hoped. The flight plan called for the two astronauts to spend the first 10 hours on the surface inside the Eagle, or Lunar Module, emerging at 2:12 A.M. Monday, E.D.T. They were to devote the time to checking the craft for any damage suffered in the landing, describing what they saw out the window, grabbing a brief snack, sleeping for four hours to rest from their fatiguing descent, eating a leisurely dinner, and then struggling into their space suits, visored helmets, boots and gloves for the EVA, or extravehicular activity, outside.

But soon after the landing, upon checking and finding the spacecraft in good condition and feeling chipper themselves, the astronauts decided to open the hatch and venture out earlier than planned. There had been speculation for days before the Apollo 11 flight began that the moon men, anxious to get to the main business at hand, might decide to hold off on their nap. And that's what they did.

"Houston," they radioed to Mission Control. "Our recommendation at this point is planning an EVA with your concurrence starting at about 8 o'clock this evening, Houston time. That is about three hours from now. . . . We will give you some time to think about that."

"Tranquillity Base [code name for the LM], Houston," came back the almost immediate reply. "We thought about it. We will support it. We'll go at that time."

Whatever the scientific factors involved, for television viewers in the United States it was a welcome switch. In the East, for example, it meant that the EVA could be seen in almost prime time Sunday night rather than the middle of the night. As it turned out, however, the astronauts' departure from the landing craft was delayed for a time when they had trouble depressurizing the cabin so that they could open the hatch. All the oxygen in the cabin had to be vented. The world waited.

Considerable time had to be spent donning the lunar spacesuit and preparing for the alien atmosphere outside. The moon has no air or water. Surface temperatures range from about 243 degrees above zero Fahrenheit in the unfiltered sunlight at lunar midday, to about 279 degrees below zero in the depths of the lunar night (the temperature when the astronauts first stepped outside was estimated at 40 to 50 degrees above; in the shadows it was 150 degrees below zero).

For these reasons, the spacesuit, with its portable life support system (P.L.S.S.), had to be almost as self-sufficient as a space ship. It carried its own supply of electricity, water and oxygen. It had a fan, a refrigeration element and a sophisticated two-way radio. In addition, it provided protection against total vacuum, the temperature extremes and the risk of puncture by a hurtling micrometeoroid. At the same time, the suit was flexible enough so that the wearer could walk, climb, dig and set out instruments on the moon's surface. Altogether, the moon-walk costume weighed 185 pounds—but it seemed like only one-sixth that weight because of the lunar gravity.

The difficulty of maneuvering in the spacesuit in close quarters was demonstrated by the following conversation between Armstrong and Aldrin as Armstrong started to back out the Eagle's hatch to descend the ladder:

Okay. Bical pump secondary circuit breaker open. Back to lean—this way. Radar circuit breakers open. Well, I'm looking head-on at it. I'll get it. Okay. My antenna's out. Right. Okay, now we're ready to hook up the LEC. Okay. Now we need to hook this. Your visor. Yep. Your back is up against the porch. Now you're clear. Over toward me. Straight down, to your left a little bit. Plenty of room. You're lined up nicely. Toward me a little bit. Down. Okay. Now you're clear. You're catching the first hinge. The what hinge? All right, move. Roll to the left. Okay, now you're clear. You're lined up on the platform. Put your left foot to the right a little bit. Okay, that's good. More left. Good. Okay. You're not quite squared away. Roll to the right a little. Now you're even. That's good. You've plenty of room to your left. It's a little close on the . . . How'm I doing? You're doing fine. Want this bag? Yeah. Got it.

Finally Armstrong announced: "Okay, Houston, I'm on the porch."

Walking Like Kangaroos

The astronauts found walking and working on the moon less taxing than had been forecast. Armstrong once reported he was "very comfortable." They seemed to have a little difficulty in adjusting their vision to the deep shadows, but their perception appeared not to suffer at all. Despite their heavy spacesuits and backpacks, the men bounded about easily in kangaroo, almost floating hops. "You do have to be rather careful to keep track of where your center of mass is," Aldrin observed after testing his agility. "Sometimes it takes about two or three paces to make sure that you've got your feet underneath you," he explained. "And about two to three, or maybe four, easy paces can bring you to a fairly smooth stop. Like a football player, you just have to put out to the side and cut a little bit. The so-called kangaroo hop—it does work, but it seems the forward ability is not quite as good as it is in the more conventional one foot after another. As far as saying what a safe pace might be—the one that I'm using now could get rather tiring after several hundred. But this may be a function of the suit as well as lack of gravity forces."

Aldrin discussed the adjustment of vision to the shadows:

"I've noticed several times in going from the sunlight into shadow that just as I go in I catch an additional reflection off the LEM that, along with reflection off my face into the visor, makes visibility very poor just at the transition of sunlight into shadow. Since we have so much glare coming onto my visor—shadow—and then it takes a short while for my eyes to adapt to the lighting conditions. Inside the shadow area, visibility is, as we said before, not too great. But with both visors up we can certainly see what sort of footprints we have and the condition of the soil. Then after being out in the sunlight a while it takes . . . watch it, Neil. Neil, you're on a cable. Yeah, lift up your right foot. Right foot. It's still hooked on it. Wait a minute. Okay, you're clear now."

It was an eerie scene, like a throwback to Buck Rogers science fiction. The black-and-white TV pictures of the bug-shaped Lunar Module and the astronauts were so sharp and clear as to seem unreal, more like a toy and toy-like figures than human beings on the most daring and far-reaching expedition thus far undertaken.

Soon after the astronauts came out of the Lunar Module, they checked the outside of the craft to determine whether any damage occurred in the landing. Here was the discussion:

ALDRIN: I say the jet deflector that's mounted on quad 4 seems to be—the surface of it—seems to be more wrinkled than the one that's on quad 1. Generally the underneath part of the LEM seems to have stood up quite well. We'll get some pictures in the aft part of the LEM that illuminate the thermal effects much better than we could get them up here at the front.

ARMSTRONG: I don't note any abnormalities in the

LEM. The pads seem to be in good shape. The primary and secondary struts are in good shape. Antennas are all in place. There's no evidence of any problem underneath the LEM due to either engine exhaust or drainage of any kind.

ALDRIN: It's very surprising, very surprising, the lack of penetration of all four of the foot pads. I'd say if we were to try and determine just how far below the surface they would have penetrated, you'd measure two or three inches, wouldn't you say, Neil?

ARMSTRONG: At the most, yes, Buzz. There is probably less than that.

ALDRIN: We need a picture of the FY strut taken from near the descent stage and I think we'd be able to see a little bit better what the thermal effects are. They seem to be quite minimal. This one picture taken at the right rear of the spacecraft looking at the skirt of the descent stage, a slight darkening of the surface color, a rather minimal amount of radiating or etching away or erosion of the surface. Now, on descent, both of us remarked that we could see a large amount of very fine dust particles moving out. It was reported beforehand that we would probably see an outgassing from the surface after the actual shutdown, but I recall I was unable to verify that. This is too big an angle, Neil.

ARMSTRONG: Yeah, I think you're right.

ALDRIN: Neil, if you'd take the camera, I'll work on the SEQ base [equipment]. Try to get some closer pictures of that rock.

ARMSTRONG: I was saying, Houston, you stop and take a photograph of something and then want to start moving again sideways, there's quite a tendency to start doing it with just gradual sideways hops.

At one point the astronauts were suddenly interrupted by a summons from Houston. Then President Nixon, to mark the momentous occasion, came on a telephone-radio hook-up to congratulate Armstrong and Aldrin in what, he said, "certainly has to be the most historic telephone call ever made." The conversation was televised at both ends and shown on a split screen, with the President in his oval office at the White House, and the astronauts standing in front of their landing craft.

"Because of what you have done," the President said, "the heavens have become a part of man's world. And as you talk to us from the Sea of Tranquillity, it requires us to redouble our efforts to bring peace and tranquillity to earth. For one priceless moment in the whole history of man, all the people on this earth are truly one—one in their pride in what you have done and one in our prayers that you will return safely to earth."

Armstrong, the Apollo 11 commander, replied: "Thank you, Mr. President. It's a great honor and privilege for us to be here representing not only the United States but men of peace of all nations, men with interests and a curiosity and men with a vision for the future."

The astronauts wasted no time settling down to their chores. Each had a checklist printed on one sleeve of his moonsuit. Aside from providing a televised impression of what the moon is like, they had two primary objectives: (1) to set out three scientific experiments; (2) to collect up to 60 pounds of lunar rocks and soil.

Because the moon, unlike the earth, is uncorrupted by the moving gases of an atmosphere and unworn by the erosive pounding of wind and water scientists hoped that the spacemen's probes would help unlock some of the geological mysteries of the solar system. Unmanned probes during the past five years had provided much information about the moon—through pictures and digging—but the major questions remained unanswered. The ability of astronauts to choose a suitable site to emplace instruments and pick out the rocks of greatest interest was one of the prime justifications for a manned landing.

For the first experiment, Armstrong and Aldrin set up a sheet of aluminum foil a foot wide and a yard long for a solar-wind test. Its purpose was to entrap rare outflowing gases from the sun—such as argon, krypton, xenon, neon and helium. The captured gases were placed in a vacuum box to be returned to earth for analysis. Scientists hoped this "wind" would throw light on the way in which the sun and planets were formed.

The second experiment was a seismometer to report any tremors caused by falling meteorites or volcanic eruptions. The reports, it was hoped, would provide clues to the composition of the moon.

The third experiment, also left on the moon, was a two-foot-square laser reflector made up of 100 fused-silica prisms. Pointed toward the earth, it was designed to reflect a beam of light directly back to the earth. By measuring the travel time of the pulses to the moon and back, scientists could use the reflector to follow subtle changes in earth-moon distances. These changes might indicate whether gravity was weakening or the continents shifting, and might provide the most sensitive tests to date of Einstein's general theory of relativity.

The astronauts had some discussion of the best location for the experiments:

ALDRIN: Have you got a good area picked out?

ARMSTRONG: Well, I think right out on the rise out there is probably as good as any. Probably stay on the higher ground there.

ALDRIN: Watch at the end of that crater. It's soft.

ARMSTRONG: Yes, that's pretty soft there, isn't it?

ALDRIN: Get a couple of close-ups on these quite rounded, large boulders.

ARMSTRONG: Yeah. About 40 feet out. I'd say to the end of that next part.

ALDRIN: It's going to be a little difficult to find a good level spot here.

ARMSTRONG: The top of that next little ridge there—wouldn't that be a pretty good place?

ALDRIN: All right. I'll put the seismometer about there.

ARMSTRONG: All right.

ALDRIN: I'm going to have to get on the other side of this rock here.

ARMSTRONG: I would go right around that to the left there. Isn't that a level spot there?

ALDRIN: I think this right here is just as level.

ARMSTRONG: Okay. [Observing the ground] Looks like salt and they have probably two per cent white minerals in 'em. White crystals. And the thing that I reported before . . . I don't believe that I believe that anymore. I think it's small craters. They look like impact craters where shot—BB shot—has hit the surface.

ALDRIN: Houston, I have the seismometer experiment over now and I'm aligning it with the sun and I'm having a little bit of difficulty getting the feed going to center. It wants to move around and around the outside.

Mining the Moon

Of all the experimental tasks, however, top priority went to the collection of rocks and soil. After years of inconclusive debate about the origin of the moon, it had become generally apparent that only with a representative collection of samples could the correct answer be determined. The samples taken by Armstrong and Aldrin, both amateur geologists with special training, were headed for the Lunar Receiving Laboratory in Houston for initial tests, and then, after a period of quarantine, to other laboratories across the country for more thorough analysis.

Armstrong scooped up the first "contingency" sample and put it in his pant-leg pocket almost as soon as he got out of the Lunar Module. This was even before Aldrin descended, and was done to assure having some lunar material if something suddenly went wrong on the surface and the astronauts had to "abort" the mission.

"Like it's a little difficult to dig through the crust," Armstrong said as he gathered up the sample. "It's very interesting. It's a very soft surface but here and there where I plug with the contingency sample collector I run into very hard surface, but it appears to be very cohesive material of the same sort."

Later, while Aldrin was busy with other tasks, Armstrong went farther from the module and collected more rocks and soil at random. Aldrin then took his turn, making a more selective collection within a radius of up to 100 feet of the craft, while Armstrong took pictures. Since the space suits, tightly inflated with oxygen, did not allow the astronauts to bend more than slightly, the men used scoops and tongs with long handles. This first lunar geological prospecting was seen very clearly on home TV screens. As the astronauts walked about, they described by radio what they saw. They put the samples in sealed aluminum boxes for the trip back to earth.

Here is how some of the conversation went:

ARMSTRONG: Now this one's right down front. And I want to know if you can see an angular rock in the foreground.

HOUSTON: Roger, we have a large angular rock in the foreground. And it looks like a much smaller rock a couple of inches to the left.

ARMSTRONG: And beyond it about 10 feet is an even larger rock that's rounded. That rock is about—the closest one to you is about—that one sticking out of the sand—about one foot. It's about a foot and a half long and it's about six inches thick. But it's standing on edge.

ALDRIN: Neil, I've got the table out . . . and the bag deployed.

HOUSTON: Roger. And we see the shadow of the LEM.

ARMSTRONG: The little hill just beyond the shadow of the LEM is a pair of elongated craters, so they appear together as 40 feet long and 20 feet across and they're probably six feet deep. We'll probably get more work in there later.

HOUSTON: Roger. And we see Buzz going about his work.

Armstrong reported that he collected about 50 pounds of soil and rock samples from several areas within the limited excursion sector around the LM. Most of the samples he scooped off the surface, but he went as deep as three inches. There was no significant change in the soil composition at that depth. He did not hit any hard bed. Armstrong said there was a wide variety of rocks, and the boulders were generally about two feet high.

The rocks were coated with surface powder, making them slippery in the deep vacuum that exists on the lunar surface. The astronauts said they found a purple rock. Some of the rocks were described as vesicular, that is, full of cavities. This is characteristic of certain forms of lava, but this does not definitely establish the rock as a lava fragment. Another rock was said to resemble biotite, a dark green or black form of mica that is characteristic of continental rocks on earth. Its presence on the moon could indicate that the history of the moon had features in common with that of the earth. However, as the astronauts pointed out, definite identification had to await return and analysis of the rocks.

As one of their last acts on the lunar surface, the moon men drove coring tubes into the surface with a hammer to capture material deep enough to be free of any exposure to exhaust gases from the rocket that lowered the LM. Aldrin said he had no trouble going in about two or three inches, but then had to pound "about as hard as I could." He drove the tube about eight or nine inches into the surface, then he noticed something puzzling.

For some reason, he said, the tube "didn't seem to want to stand upright. I'd keep driving it in and it would dig some sort of a hole, but it wouldn't penetrate in a way that it would support itself." He added that the material in the tube was "quite well packed, a good bit darker and the way it adhered to the core tube gave me the distinct impression of being moist."

All in all, man's first walk on the moon lasted two hours and 21 minutes. When they completed their assignments, Armstrong and Aldrin climbed back up into the Eagle where they continued to radio their impressions to Houston before resting.

As a final description of the landing site, Armstrong reported the following: "We are landed in a relatively clear crater field . . . of circular secondary craters, most of which have rims irrespective of their rays and irrespective of their size. There are a few of the smaller craters around which do not have a discernible rim. The ground mass throughout the area is a very fine sand to a silt. I say the thing that would be most like it on earth is powdered graphite."

Successful completion of the extravehicular activity, of course, brought congratulations from around the world and put gleams in the eyes of scientists waiting eagerly for the lunar specimens. But space officials, ever mindful of the complexities of space flight, cautioned about premature cheers. Armstrong and Aldrin still had the risky lift-off to rejoin the orbiting Columbia command ship which Colonel Collins had been piloting all this time. It would be the first launching without the benefit of all the familiar accoutrements, such as concrete bases and steel gantries. Moreover, the two astronauts were completely dependent upon the 3,500-pound-thrust ascent rocket in the upper half of the Lunar Module. If it failed to fire, they would be stranded.

Chapter XXI

Home to Earth

Suspense on a space mission is always greatest when the ship is out of ground contact or is trying a maneuver for the first time. The suspense was never greater than on Monday, July 21, when Armstrong and Aldrin prepared to lift off the moon for the return home.

The ascent rocket in the upper half of the Lunar Module had been tested hundreds of times, of course; but never on the moon. And instead of a fully equipped launching pad for the lift-off, the astronauts had to rely on the LM's lower half, which had been separated from the ascent stage by the firing of explosive bolts. At this point, the mission and the men's lives depended on these two pieces of equipment working perfectly.

"You're cleared for takeoff," Mission Control radioed the astronauts. "Roger, understand. We're Number One on the runway," replied Armstrong.

The engine fired on time and at full thrust. It lifted the ascent vehicle and sent it into a long orbital path toward a rendezvous with the waiting Command Ship. A sigh of relief went up at Mission Control in Houston as the LM radio reported: "Nine, eight, seven, six, five, first stage engine on ascent. Proceed. Beautiful, 26, 36 feet per second up. Little pitch over, very smooth, very quiet ride."

Four minutes later, the Eagle reported: "Eagle is back in orbit, having left Tranquillity Base, and leaving behind a replica from our Apollo 11 patch with an olive branch."

That's not all the moon explorers left behind. On the site of the landing was the forlorn descent stage, no longer needed and therefore excess baggage. Around it, like the clutter left by some campers, were the cameras, walking boots, equipment boxes, an aluminum pole, two back packs and other equipment which the astronauts pitched out to lighten their load for take-off.

The lift-off at 1:55 P.M., E.D.T., came as Collins was piloting the Command Ship through its 25th revolution of the moon. Collins, who was somewhat of a forgotten man during the time his crew mates were performing their feats on the lunar surface, received an unofficial tribute from the space agency for doing the tasks usually shared by all three crewmen. The announcer at Mission Control said: "Not since Adam has any human known such solitude as Mike Collins is experiencing during the 47 minutes of each lunar revolution when he's behind the moon with no one to talk to except his tape recorder aboard Columbia."

Armstrong and Aldrin closed in for the rendezvous in a series of four maneuvers with the Lunar Module's smaller thruster jets. These maneuvers raised and circularized the vehicle's orbit, then slowed its flight to ease in close to the Columbia. At 5:35 P.M., some 69 miles above the front face of the moon, the Columbia joined its nose with the top hatch of the Eagle.

"That was a funny one," Collins radioed to Mission Control. "You know, I didn't feel us touch, and I thought things were pretty steady . . . and that's when all hell broke loose." As the two vehicles moved in for the link-up, their aim was off slightly; but the docking, though bumpy, was accomplished without trouble.

Armstrong and Aldrin crawled through the connecting tunnel to rejoin Collins, and then they jettisoned the Lunar Module which became a piece of junk orbiting the moon. Finally, they fired the Apollo's main engine, boosting the craft out of lunar orbit for the 60-hour coasting voyage toward the splashdown in the Pacific.

"Open up the LRL doors, Charlie," radioed Armstrong, alluding to the Lunar Receiving Laboratory in Houston where the astronauts were to remain in post-flight quarantine and the lunar samples they were bringing back were to be analyzed.

Routine Trip Back

The trip back was routine. Until the last few hours before splashdown, the astronauts had little work to do except for monitoring spacecraft systems, performing housekeeping chores, and conducting more television shows for the folks at home.

On one telecast Aldrin demonstrated how easy it was to apply ham spread to a slice of bread, which he grabbed out of mid-air in the cabin. Collins took a playful drink by holding the water gun a few inches from his mouth and squirting the water in, much as a Spaniard drinking from a wineskin. Armstrong observed: "No matter where you travel, it is nice to get home."

On another telecast the astronauts reflected movingly on what they had been through. Collins said, "This trip of ours to the moon may have looked simple and easy. I want to assure you that this has not been the case." He pointed out all the complex equipment involved, and cited the thousands of workers "below the surface" who made the mission possible.

"We've come to the conclusion," Aldrin added, "that this has been far more than three men on a voyage to the moon, more still than the efforts of a government and industry team, more even than the efforts of one nation. We feel that this stands as a symbol of the insatiable curiosity of all mankind to explore the unknown."

Although Apollo Project officials cautioned that the flight was not over until splashdown, they could hardly contain their relief, pride and optimism as the spacecraft sped earthward. Lieut. Gen. Samuel C. Phillips, the Apollo program director, told a news conference: "The men and equipment that are Apollo 11 have performed to perfection. Perfection is not too strong a word."

And George E. Mueller, the space agency's associate administrator, began reflecting on the meaning of Apollo's achievement. "It seems quite clear," he said, "that the planets of the solar system are well within our ability to explore both manned and unmanned at the present time."

It was on their trip back that the astronauts learned that the Soviet Union's unmanned spacecraft, Luna 15, which had held the world in suspense for more than a week, had apparently crashed into the moon's surface. Launched three days ahead of Apollo 11, Luna 15 was regarded by many observers as a Soviet rival to the American spacecraft. The Russians announced Tuesday morning that Luna 15 had "reached the moon's surface" and that its work had "ended." The fact that it went out of action as soon as it touched the lunar surface suggested that it was not a soft landing but a crash.

As a precaution for Apollo 11's splashdown, flight planners changed the target area because of worsening weather conditions near the original site. Officials were reluctant to gamble on bringing the spacecraft down into unpredictable and potentially stormy conditions that might hamper the rescue helicopters or produce such heavy swells that the astronauts could become seasick upon landing. The change posed no particularly vexing problems for the crew.

At 12:22 P.M. Thursday, the astronauts jettisoned the Service Module, the equipment section carrying the rocket engine, to start the final descent from 400,000 feet into the earth's atmosphere. When they reached the atmosphere, the earth's gravity was exerting such a pulling force that the capsule was traveling at more than 24,000 miles an hour. Then the earth's air gripped at the spacecraft, slowing its speed.

Safely Home

Within minutes, the astronauts were in the water 950 miles southwest of Hawaii—home safe from man's first trip to the moon. The splashdown was one minute earlier than the time scheduled in the original flight plan. Waiting to welcome them aboard the recovery aircraft carrier, the U.S.S. *Hornet,* 11 miles away, was President Nixon. He was in the Pacific area en route to a tour of Asian countries.

"All three of us are excellent," the Apollo men radioed from their bobbing capsule. "Take your time."

The astronauts were given biological isolation garments by a frogman as soon as the hatch was open, as part of the quarantine program designed to avoid contamination by possible lunar organisms. And they were washed off when they climbed into a raft to await the recovery helicopter.

President Nixon watched the recovery proceedings from the *Hornet*'s bridge with evident relish, but he had to wait 55 minutes to talk with them after they reached the deck of the carrier. The helicopter that brought them was immediately towed to an elevator for lowering to the flight deck where the astronauts went into the mobile isolation trailer for a quick checkup.

When the meeting finally came, the smiling astronauts were peeking out a window of the van. Now wearing blue coveralls, the men appeared relaxed and happy as they jostled each other to look out the window. The President, with the band playing "Hail to the Chief," strode briskly down the carpet to the front of the van. He stood about three feet from the astronauts and spoke to them through a microphone-speaker hook-up:

"Neil, Buzz and Mike, I want you to know that I think I'm the luckiest man in the world, and I say this not only because I have the honor to be President of the United States, but particularly because I have the privilege of speaking for so many in welcoming you back to earth.

"I can tell you about all the messages we have received in Washington, over 100 foreign governments, emperors and presidents and prime ministers and kings, have sent the most warm messages that we've ever received.

"They represent over two billion people on this earth, all of them who have had the opportunity to see what you have done."

As throngs of sailors listened on the deck and millions of people watched on home television screens, Mr. Nixon then chatted with the astronauts about baseball, world news, the splashdown and phone calls he made to their wives, "three of the greatest ladies," inviting them and the astronauts to a state dinner. In the middle of the banter, he suddenly blurted, "Gee, you look great! You feel as good as you look?"

"I feel just perfect, Mr. President," Armstrong, the Apollo commander, replied.

Then the President grew serious again:

"I was thinking as you know as you came down and we knew it was a success and it had only been eight days—just a week, a long week—but this is the greatest week in the history of the world since the Creation. Because of what happened in this week, the world is bigger infinitely and also as I'm going to find in this trip around the world, and Secretary [of State William] Rogers will find as he covers the other countries in Asia, as a result of what you've done, the world's never been closer together before.

"And we just thank you for that, and I only hope that all of us in government, all of us in America, that as a result of what you've done we could do our job a little better, we can reach for the stars just as you have reached so far for the stars."

Armstrong, again responding for the crew, said: "We're just pleased to be back and very honored that you were so kind as to come out here and welcome us back, and we look forward to getting out of this quarantine and talking."

At the end of the brief ceremony, the President waved to them and walked back to return to the deck. As he left, the astronauts waved out the window at the cheering sailors. Then they, too, turned into their isolation van to continue their quarantine.

The astronauts were to spend three days in the mobile trailer while it was shipped by the *Hornet* to Hawaii, thence by plane and truck to the Lunar Receiving Laboratory in Houston. There they remained in quarantine until August 11 undergoing extensive post-flight debriefings. The 50 pounds of lunar rocks and soil they brought back were shipped immediately to the lab for analysis.

"Task Accomplished"

The completion of the epic mission triggered rousing celebrations across the nation, particularly at the control centers and at the homes of the astronauts' families. At Mission Control in Houston, the men stood up at their consoles, cheered, waved small American flags and talked of flying men to Mars. On a small screen there appeared the Eagle emblem of the Apollo 11 and the words: "Task Accomplished. July 1969."

"Many of us still can't believe that the goal we set out to achieve in 1961 has been achieved," commented George M. Low, the manager of the Apollo spacecraft program.

The wives and children of the returning astronauts were happy, relieved and proud. Mrs. Armstrong said, "If anyone were to ask me how I could describe this flight, I can only say it was out of this world." At the Aldrin home, where fireworks went off on the front lawn when the splashdown was passed, Mrs. Aldrin said she had offered "just one big prayer during the mission." Mrs. Collins added that "the whole thing was marvelous, marvelous."

There were a few skeptics. In a tavern in Madison, Wisconsin, one of them declared: "This is the greatest hoax ever perpetrated on mankind. Those guys were in Nevada the whole time and never got more than 30 feet off the ground." There was an echo in Chicago, where a restaurant patron informed a waitress: "Those guys never walked on the moon. It was one of those Hollywood tricks."

But there were no doubters in Wapakoneta, Ohio, Armstrong's hometown. The high school band marched triumphantly down Blackhoof Street playing "Moon over Miami" and "Harvest Moon." And the town fathers were already talking about opening an Armstrong Museum.

There were cheers, yes, but also sober thoughts as scientists, philosophers and journalists began speculating about the future, the meaning of the Apollo achievement and the perspective of man. Perhaps representative of some of this speculation was this reflection by James Reston of *The New York Times*:

"The great achievement of the men on the moon is not only that they made history, but that they expanded man's vision of what history might be. One moon landing doesn't make a new heaven and a new earth, but it has dramatized the possibilities of doing so."

Epilogue

Man will never again look up at the moon and see it as the silvery symbol of all that is impossible.

Man may now see the moon as a new world to explore or even colonize, a new frontier for his restless, curious mind. One long voyage ended will only begin a second. There may be setbacks. The conquest of each new frontier has taken its toll. But there will surely be many more landings on the moon, at different places and for longer durations. There will eventually be laboratories orbiting the moon, giant telescopes peering at the heavens with airless clarity and permanent outposts in domed villages on the surface. "It will be like Phoenix," Thomas O. Paine, the NASA administrator, has said. "Brilliant sun and stars, dramatic landscapes, mountains and craters, a magnificent view of earth. The loneliness so lonely, the togetherness so together. It will be a cruise ship and a prison, Hawaii and Alaska. Don't forget, this will take place in a time of the four-day work week, of general affluence. People will be free to carry out activities that suit their life style. They will be actively seeking excitement, even danger, a sense of striving together in a great enterprise."

Man may now see the moon as a steppingstone to the distant planets. It may become a place from where, because of its weak gravity, the more far-reaching expeditions of the future will be launched. Laboratories established there may be prototypes of those set up later on Mars, for it is not unreasonable to think of men walking on that planet before the century is out. "While the moon has been the focus of our efforts," Paine said in 1969, "the true goal is far more than being first to land men on the moon, as though it were a celestial Mount Everest to be climbed. The real goal is to develop and demonstrate the capability for interplanetary travel."

Man may now also see the moon and, knowing that he

has been there, learn something about himself. He could gain a new sense of what is possible, a transfusion of confidence in himself. Considering that, in the 1960s, man dwelled so despairingly on his failures and limitations, such a change in the human attitude may be the most significant result of the Apollo voyage to the moon.

Is this too much to hope for?

The answer—the true assessment of Apollo's significance —may not be forthcoming for decades or centuries. Columbus could not see beyond the gold and spices he sought. Neither can we see with any assurance beyond the prestige, power and scientific knowledge that were the treasures sought for their nation by Armstrong, Aldrin and Collins.

But through knowledge man's perspectives do change. The great seafaring explorers broadened man's conception of his world. Copernican astronomy disabused man of the medieval conceit that he was at the center of everything. Darwin changed man's conceptions of his origins. Freud gave man new insights into the emotions that shape his life.

To reach the moon, to see the earth as a small planet— such extensions of man's dominion over nature cannot help but have an impact on his image of himself and of what he can do. This is the hope and promise of man's first voyage to the moon. For Apollo, whatever its place in history, showed what remarkable achievements a society can accomplish, given adequate leadership, national resolve and personal courage.

Appendixes

Log of Conversations and Announcements
July 20-21, 1969

*Excerpts from Radio Communications as Recorded
and Transcribed by* The New York Times.
All times are Eastern daylight.

**Sunday, July 20—as Lunar Module descended to surface
of moon**

EAGLE (12:57 P.M.): Undocked.

HOUSTON: Roger. How does it look?

EAGLE: The Eagle has wings.

HOUSTON: Roger.

EAGLE: Looking good.

HOUSTON: Roger, Mike, my mark seven minutes to
ignition. Mark. Seven minutes.

COLUMBIA: Roger.

HOUSTON: You're looking good for separation. You
are go for separation, Columbia. Over.

COLUMBIA: Okay. Eagle one minute to take. Take
care.

EAGLE: See you later.

COLUMBIA: You've got a fine looking flying machine
there, Eagle, despite the fact you're upside down.

EAGLE: Somebody's upside down.

* * *

EAGLE (2:14 P.M.): We're going right down U.S. 1,
Mike.

* * *

APOLLO CONTROL (3:43 P.M.) at 102 hours 12
minutes into the flight of Apollo 11: We're now 2 minutes

53 seconds from reacquiring the spacecraft, 21 minutes 23 seconds from the beginning of the powered descent to the lunar surface. It's grown quite quiet here in Mission Control. A few moments ago, Flight Director Gene Kranz requested that everyone sit down, get prepared for events that are coming and he closed with the remark: Good luck to all of you. Now here on the front of our display boards, we have a number of big plot boards which will be used to keep track of the burn progress. Among the more important of those is one which will show the performance of onboard guidance systems both the primary and the backup guidance system and compare the guidance systems with the manned space flight network tracking. These displays by the time this is all over will look a good deal like a combination Christmas tree-Fourth of July.

We're now 1 minute 39 seconds from reacquiring the command module Columbia. Acquisition of the lunar module will come a little less than 2 minutes after that. At the time we acquire the LM—that should be at an altitude of about 18 nautical miles descending toward the 50,000-foot pericynthion [low point of moon orbit] from which point the power descent to the lunar surface will be initiated.

If for any reason the crew does not like the way things look as they are coming across the pericynthion simply by not initiating the maneuver they will remain in a safe orbit of 60 miles by 50,000 feet and if they desire they would be able to attempt the power descent on a following revolution on a ground elapsed time of about 104 hours 26 minutes.

We are now coming up on 30 seconds to acquisition of the command module. We'll stand by for that event.

*   *   *

APOLLO CONTROL: We're now in the approach phase. Everything looking good. Altitude 5,200 feet.

EAGLE: Manual attitude control is good.

APOLLO CONTROL: Altitude 4,200.

HOUSTON: You are go for landing. Over.

EAGLE: Roger go for landing, 3,000 feet. We're go. We're go. Two thousand feet. Two thousand feet. Into the AGS 47 degrees.

HOUSTON: Roger.

HOUSTON: Eagle, looking great. You're go.

APOLLO CONTROL: Altitude 1,600 . . . 1,400 feet, still looking very good.

EAGLE: 35 degrees. 35 degrees 750 coming down to 23; 700 feet 21 down, 33 degrees; 600 feet down to 19; 540 feet down to 15; 400 feet down at 9 three forward;

350 feet down at 4; 300 feet down 3½; 47 forward; on one a minute 1½ down; 270 . . . 50 down at 2½, 19 forward. Altitude velocity lights down 15 forward, 11 forward; 200 feet 4½ down 5½ down 6½ down, 5½ down, nine forward; 120 feet 100 feet 3½ down, nine forward, 5 per cent 75 feet looking good down ½, six forward.

HOUSTON: Sixty seconds.

EAGLE: Lights on; down 2½. Forward, forward 40 feet down 2½ picking up some dust; 30 feet 2½ down shadow, four forward, four forward, drifting to the right a little.

HOUSTON: Thirty seconds.

EAGLE: Contact light. Okay, engines stop. Engine arm off.

HOUSTON: We copy. You're down, Eagle.

EAGLE: Houston. Tranquility Base here. The Eagle has landed.

Sunday, July 20—from the lunar surface

HOUSTON: Roger, Tranquility, we copy you on the ground. You've got a bunch of guys about to turn blue. We're breathing again. Thanks a lot.

TRANQUILITY BASE: Thank you.

HOUSTON: You're looking good here.

TRANQUILITY BASE: A very smooth touchdown.

HOUSTON: Eagle, you are stay for T1. [The first step in the lunar operation.] Over.

TRANQUILITY BASE: Roger. Stay for T1.

HOUSTON: Roger and we see you venting the ox.

TRANQUILITY BASE: Roger.

COLUMBIA: How do you read me?

HOUSTON: Columbia, he has landed Tranquility Base. Eagle is at Tranquility. I read you five by. Over.

COLUMBIA: Yes, I heard the whole thing.

HOUSTON: Well, it's a good show.

COLUMBIA: Fantastic.

TRANQUILITY BASE: I'll second that.

* * *

HOUSTON: Eagle, you are stay for T2. Over.

TRANQUILITY BASE: Roger. Stay for T2. We thank you.

HOUSTON: Roger, sir.

APOLLO CONTROL: That's stay for another two minutes plus. The next stay-no stay will be for one revolution.

TRANQUILITY BASE: Houston, that may have seemed

like a very long final phase but the auto targeting was taking us right into a football field-sized crater with a large number of big boulders and rocks for about one or two crater diameters around it. And it required us to fly manually over the rock field to find a reasonably good area.

HOUSTON: Roger. We copy. It was beautiful from here, Tranquility. Over.

TRANQUILITY BASE: We'll get to the details of what's around here but it looks like a collection of just about every variety of shape, angularity, granularity, about every variety of rock you could find. The colors vary pretty much depending on how you are looking relative to the zero phase length. There doesn't appear to be too much of a general color at all. However, it looks as though some of the rocks and boulders, of which there are quite a few in the near area—it looks as though they're going to have some interesting colors to them. Over.

HOUSTON: Roger. Copy. Sounds good to us, Tranquility. We'll let you press on through the simulated countdown and we'll talk to you later. Over.

TRANQUILITY BASE: Okay, this one-sixth G is just like an airplane.

HOUSTON: Roger, Tranquility. Be advised there are lots of smiling faces in this room and all over the world. Over.

TRANQUILITY BASE: There are two of them up here.

HOUSTON: Roger. It was a beautiful job, you guys.

COLUMBIA: And don't forget one in the command module.

TRANQUILITY BASE: Roger.

* * *

COLUMBIA: Do you have any idea whether they landed left or right of center line—just a little bit long. Is that all we know?

HOUSTON: Apparently that's about all we can tell. Over.

COLUMBIA: Okay, thank you.

* * *

TRANQUILITY BASE: Okay. I'd say the color of the local surface is very comparable to that we abserved from orbit at this sun angle—about 10 degrees sun angle or that nature. It's pretty much without color. It's gray and it's very white as you look into the zero phase line. And it's considerably darker gray, more like an ashen gray, as you look out 90 degrees to the sun. Some of the surface rocks in close here that have been fractured or disturbed by the rocket engine plume are coated with this light gray on the outside. But where they've been broken, they display a dark, very dark, gray interior and it looks like it could be country basalt.

*   *   *

HOUSTON: Columbia, Houston. Two minutes to LOS [loss of signal]. You're looking great. Going over the hill. Over.

COLUMBIA: Okay. Thank you. Glad to hear it's looking good. Do you have a suggested attitude for me? This one here seems all right.

HOUSTON: Stand by.

COLUMBIA: Let me know when it's lunch time, will ya?

HOUSTON: Say again?

HOUSTON: Columbia, Houston. You got a good attitude right there.

APOLLO CONTROL: This is Apollo Control. We've had loss of signal now from the command module. Of course, we'll maintain constant communication with the lunar module on the lunar surface. We have some heart rates for Neil Armstrong during that powered descent to lunar surface. At the time the burn was initiated, Armstrong's heart rate was 110. At touchdown on the lunar surface, he had a heart rate of 156 beats per minute, and the flight surgeon reports that his heart rate is now in the 90's. We do not have biomedical data on Buzz Aldrin.

*   *   *

APOLLO CONTROL: Ladies and gentlemen, I'd like to at this time introduce the administrator of the National Aeronautics and Space Administration, Dr. Thomas O. Paine. I have a short statement then we'll be glad to accept questions. Dr. Paine.

DR. PAINE: Immediately after the lunar touchdown I called the White House from Mission Control and gave the following report to the President:

Mr. President, it is my honor on behalf of the entire NASA team to report to you that the Eagle has landed on the Sea of Tranquility and our astronauts are safe and looking forward to starting the exploration of the moon. We then discussed the gripping excitement and wonder that has been present in the White House and in Mission Control during the final minutes of this historic touchdown. I emphasized to the President the fact that we still had many difficult steps ahead of us in the Apollo 11 mission, but that at the same time a giant step had been made with our successful landing.

President Nixon asked me to convey to all of the NASA team and its associated industrial and university associates his personal congratulations on the success of the initial lunar landing and gave us his good wishes for the continuing success of this mission.

* * *

HOUSTON: Tranquility Base, the white team is going off now and the maroon team take over. We appreciate the great show; it was a beautiful job, you guys.

TRANQUILITY BASE: Roger. Couldn't ask for better treatment from all the way back there.

TRANQUILITY BASE: Houston, our recommendation at this point is planning an EVA [extra-vehicular activity] with your concurrence starting at about 8 o'clock this evening, Houston time. That is about three hours from now.

HOUSTON: Stand by.

TRANQUILITY BASE: We will give you some time to think about that.

HOUSTON: Tranquility Base, Houston. We thought about it. We will support it. We'll go at that time.

TRANQUILITY BASE: Roger.

HOUSTON: You guys are getting prime time on TV there.

TRANQUILITY BASE: I hope that little TV set works. We'll see.

* * *

TRANQUILITY BASE: This is the LM pilot. I'd like to take this opportunity to ask every person listening in, whoever, wherever they may be, to pause for a moment and contemplate the events of the past few hours and to give thanks in his or her own way. Over.

HOUSTON: Roger, Tranquility Base.

APOLLO CONTROL (7:15 P.M.): You heard that statement in our tapes transmission from lunar module pilot Buzz Aldrin. Our projected time for Extra Vehicular Activity at this point is still very preliminary. I repeat, it could come as soon as 8 P.M., Houston time. We won't know for sure about the time with reasonable certainty until about an hour before the event. Meanwhile, we'll soon be progressing toward man's first step on the lunar surface. We have an interesting phenomena here in the Mission Control Center, Houston, something that we've never seen before. Our visual of the lunar module—our visual display now standing still, our velocity digitals for our Tranquility base now reading zero. Reverting, if we could, to the terminology of an earlier form of transportation—the railroad—what we're witnessing now is man's very first trip into space with a station-stop along the route.

HOUSTON: Tranquility Base, Houston. We'd like some

estimate of how far along you are with your eating and when you may be ready to start your EVA prep. Over.

TRANQUILITY BASE: I think that we'll be ready to start EVA prep in about a half hour or so.

HOUSTON: Roger, Tranquility.

*   *   *

HOUSTON (8:17 P.M.): Columbia, go ahead.

COLUMBIA: One of the craters I was talking about is located exactly at 56.7.

HOUSTON: Roger, we found that one.

COLUMBIA: The other one's located at 7.2 two-thirds of the way from . . .

HOUSTON: Roger, we believe you were looking a little too far to the west and south.

COLUMBIA: Roger, I was looking where . . . was tracking on the average and I understand it should have been more to the north and more to the west; actually, a tiny bit outside the circle.

HOUSTON: More to the north and a little more to the east. The feature that I was describing to you, the small bright crater on the rim of the large fairly old crater, would be about Mike .8 and 8.2.

HOUSTON: Tranquility Base, this is Houston. Can you give us some idea where you are in the surface checklist at the present time.

TRANQUILITY BASE: They were at the top of page 27.

COLUMBIA: Roger. Finally got you back on. I've been unsuccessfully trying to get you on the high gain and I've got command to reset the process. How do you read me now?

HOUSTON: Roger. I hear you loud with background noise.

COLUMBIA: Omni Delta and you were cut out and I never got your coordinance or estimated LM position.

HOUSTON: Estimated LM position is latitude plus .799, longitude over 2 plus 11.730.

COLUMBIA: What I'm interested in is direct coordinance on that map reading.

COLUMBIA: Could you enable the S-band relay at least one way from Eagle to Columbia, so I can hear what's going on?

HOUSTON: Roger. There's not much going on at the present time, Columbia. I'll see what I can do about the relay. . . .

HOUSTON: Columbia, this is Houston. Are you aware that Eagle plans the EVA about four hours early?

COLUMBIA: Affirmative. I haven't had any word from those guys and I thought I'd be hearing them through your S-band relay.

APOLLO CONTROL (8:48 P.M.): We'll still have acquisition of Columbia for another eight minutes. All systems in Eagle still looking good. Cabin pressure 4.86 pounds, showing a temperature of 63 degrees in the Eagle's cabin.

COLUMBIA: During the next pass I'd appreciate the S-band relay mode.

HOUSTON: We're working on that. There haven't been any transmissions from Tranquility Base since we last talked to you.

APOLLO CONTROL: We've had loss of signal on Columbia. The clock here at Control Center counting down to depressurization time on Eagle shows we're 36 minutes, 39 seconds away from that event. We believe the crew is pretty well on the time line in the EVA preparation.

APOLLO CONTROL (9:36 P.M.): This latest report the crew is—they're getting the electrical checkout—indicates they are about 40 minutes behind the time line. We will acquire Columbia in six minutes.

TRANQUILITY BASE: How do you read now?

HOUSTON: Okay. I think that's going to better.

HOUSTON: We have acquisition of Columbia.

HOUSTON: Roger, Columbia. Reading you loud and clear on the high gain. We have enabled the one-way Nixon relay that you requested. The crew of Tranquility Base is currently donning PLSSes [portable life support systems]. Com checks out.

COLUMBIA: Sounds okay.

TRANQUILITY BASE (9:45 P.M.): Houston, Tranquility. You'll find that the area around the ladder is in a complete dark shadow, so we're going to have some problem with TV. But I'm sure you'll see the—you'll get a picture from the lighted horizon.

HOUSTON: Neil, Neil, this is Houston. I can hear you trying to transmit. However, your transmission is breaking up.

TRANQUILITY BASE: Neil's got his antenna up now. Let's see if he comes through any better now.

TRANQUILITY BASE: Okay, Houston, this is Neil. How do you read?

HOUSTON: Neil, this is Houston, reading you beautifully.

TRANQUILITY BASE: My antenna's scratching the roof. Do we have a go for cabin depress?

COLUMBIA: They hear everything but that.

TRANQUILITY BASE: Houston, this is Tranquility. We're standing by for go for cabin depress.

HOUSTON: You are go for cabin depressurization. Go for cabin depressurization.

COLUMBIA (10 P.M.): I don't know if you guys can read me on VHF, but you sure sound good down there.

TRANQUILITY BASE: Okay, the vent window is clear. I remove lever from the engine cover.

HOUSTON: Buzz, you're coming through loud and clear, and Mike passes on the word that he's receiving you and following your progress with interest.

TRANQUILITY BASE: Lock system, decks, exit check, blue locks are checked, lock locks, red locks, perch locks, and on this side the perch locks and lock locks—both sides, body locks, and the calm.

HOUSTON (10:17 P.M.): Columbia, this is Houston. Do you read?

COLUMBIA: Read you loud and clear.

HOUSTON: Were you successful in spotting the LM on that pass?

COLUMBIA: Negative. I checked both locations and it's no dice.

APOLLO CONTROL (10:25 P.M.): In the control center a clock has been set up to record the operating time on Neil Armstrong's total life support system. EVA will be counted from that time.

TRANQUILITY BASE: Cabin repress closed. Now comes the gymnastics. Air pressure going toward zero. Standby LM suit circuit 36 to 43. That's verified. FIT GA pressure about 4.5, 4.75 and coming down. We'll open the hatch when we get to zero. Do you want to bring down one of your visors now or leave them up? We can put them down if we need them. We have visor down.

APOLLO CONTROL (10:33 P.M.): Coming up on five minutes of operation of Neil Armstrong's portable life support system now.

HOUSTON (10:37): Neil, this is Houston, what's your status on hatch opening?

EAGLE: Everything is go here. We're just waiting for the cabin pressure to bleed to a low enough pressure to open the hatch. It's about .1 on our gauge now.

ALDRIN: I'd hate to tug on that thing. Alternative would be to open that one too.

HOUSTON: Neil, this is Houston. Over.

EAGLE: We're going to try it.

HOUSTON: Roger.

EAGLE: The hatch is coming open.

ALDRIN: Hold it from going closed and I'll get the valve turner. I'd better get up first.

       *   *

ARMSTRONG: Okay Houston, I'm on the porch.

HOUSTON: Roger, Neil.

ALDRIN: Halt where you are a minute, Neil.

ARMSTRONG: Okay.

APOLLO CONTROL: Neil Armstrong on the porch at 109 hours 19 minutes 16 seconds.

       *   *

HOUSTON: Roger. We're getting a picture on the TV.

ALDRIN: You've got a good picture, huh?

HOUSTON: There's a great deal of contrast in it and currently it's upside down on monitor. But we can make out a fair amount of detail.

EAGLE: Okay, will you verify the position, the opening I ought to have on the camera.

HOUSTON: The what? Okay, Neil. We can see you coming down the ladder now.

ARMSTRONG: Okay. I just checked getting back up to that first step. It didn't collapse too far. But it's adequate to get back up.

HOUSTON: Roger. We copy.

ARMSTRONG: It's a pretty good little jump.

HOUSTON: Buzz, this is Houston. F2 1/160th of a second for shadow photography on the sequence camera.

ARMSTRONG: I'm at the foot of the ladder. The LM foot beds are only depressed in the surface about one or two inches, although the surface appears to be very, very fine grained as you get close to it. It's almost like a powder. It's very fine. I'm going to step off the LM now.

That's one small step for a man, one giant leap for mankind.

The surface is fine and powdery. I can pick it up loosely with my toe. It does adhere in fine layers like powdered charcoal to the sole and the sides of my boots. I only go in a small fraction of an inch, maybe an eighth of an inch but I can see the footprints of my boots and the treads in the fine sandy particles.

There seems to be no difficulty in moving around this and we suspect that it's even perhaps easier than the simulations of 1/6 G that we performed in various simulations on the ground. Actually no trouble to walk around. The descent engine did not leave a crater of any size. It has about one-foot clearance on the ground. We're essentially on a very level place here. I can see some evidence of rays emanating from the descent engine, but a very insignificant amount.

*   *   *

HOUSTON: Neil, this is Houston did you copy about the contingency sample? Over.

ARMSTRONG: Roger. Going to get to that just as soon as I finish these picture series.

ALDRIN: Are you going to get the contingency sample? Okay. That's good.

ARMSTRONG: The contingency sample is down and it's up. Like it's a little difficult to dig through the crust. It's very interesting. It's a very soft surface but here and there where I plug with the contingency sample collector I run into very hard surface but it appears to be very cohesive material of the same sort. I'll try to get a rock in here.

HOUSTON: Oh, that looks beautiful from here, Neil.

ARMSTRONG: It has a stark beauty all its own. It's like much of the high desert of the United States. It's different but it's very pretty out here. Be advised that a lot of the rock samples out here, the hard rock samples have what appear to be vesicles in the surface. Also, as I look at one now that appears to have some sort of feenacres [spelled phonetically].

HOUSTON: Roger. Out.

ARMSTRONG: This has been about six or eight inches into the surface. It's easy to push on it. I'm sure I could push it in farther but it's hard for me to bend down farther than that.

ALDRIN: I didn't know you could throw so far, Neil.

ARMSTRONG: See me throw things? Is my pocket open?

ALDRIN: Yes it is. It's not up against your suit, though. Hit it back once more. More toward the inside. Okay that's good.

ARMSTRONG: Put it in the pocket.

ALDRIN: Yes. Push down. Got it? No it's not all the way in. Push. There you go.

ARMSTRONG: The sample is in the pocket. My oxygen is 81 percent. I have no flags and I'm in minimum flow.

HOUSTON: Roger, Neil.

*   *   *

ALDRIN: How far are my feet from the . . .

ARMSTRONG: O.K., you're right at the edge of the porch.

*   *   *

ALDRIN: Now I want to back up and partially close the hatch—making sure not to lock it on my way out.

ARMSTRONG: . . . Good thought.

ALDRIN: That's our home for the next couple hours; we want to take good care of it. O.K., I'm on the top step and I can look down over the ICU and landing gear pad. It's a very simple matter to hop down from one step to the next.

ARMSTRONG: Yes, I found that to be very comfortable, and walking is also very comfortable, Houston.

ARMSTRONG: You've got three more steps and then a long one.

ALDRIN: O.K. I'm going to leave that one foot up there and both hands down to about the fourth rung up.

ARMSTRONG: Now I think I'll do the same.

A little more. About another inch. There you got it. That's a good step.

About a three footer.

Beautiful view.

Ain't that somethin'?

*   *   *

ARMSTRONG: Isn't it fun?

ALDRIN: Right in this area I don't think there's much of . . . fine powder . . . It's hard to tell whether it's a clod or a rock.

ARMSTRONG: You can pick it up—

ALDRIN: And it bounces.

ALDRIN: Reaching down is fairly easy. The mass of the backpack does have some effect in inertia. There's a slight tendency I can see now to backwards due to the soft, very soft, texture.

ARMSTRONG: You're standing on a big rock there now.

*   *   *

HOUSTON: Neil, this is Houston. That's affirmative— we're getting a new picture. You can tell it's a longer focal length lens and for your information all LEM systems are go. Over.

EAGLE: I appreciate that. Thank you. Bill is now unveiling the plaque that is. . . .

HOUSTON: Roger, we've got you foresighted, but back to one side.

EAGLE: For those who haven't read the plaque, we'll read the plaque that's on the front landing gear of this LEM. This is two hemispheres—one showing each of the two hemispheres of earth MN—underneath it says "Where men from the planet Earth first set foot upon the moon, July

1969, A.D. They came in peace for all mankind." It has the crew members' signatures and the signature of the President of the United States.

ALDRIN: I'm afraid those materials are going to get dusty. The surface material is powdery no matter how good your lens is but if you got smudges. It's very much like a very finely powdered carbon. But it is pretty looking.

ARMSTRONG: Would you pull out some of my cable for me, Buzz?

ALDRIN: Houston, how close are you able to get things in focus?

HOUSTON: This is Houston. We can see Buzz's right hand. It's somewhat out of focus. I'd say we were focussing down to probably oh, eight inches to a foot behind the position of his hand when he was pulling out the cable.

*   *   *

HOUSTON: Columbia, this is Houston reading you loud and clear. Over.

COLUMBIA: I read you loud and clear—how's it going?

HOUSTON: Roger, the EVA is progressing beautifully. I believe they're setting up the flag now.

COLUMBIA: Great!

HOUSTON: I guess you're about the only person around that doesn't have TV coverage of the scene.

COLUMBIA: That's right—I don't mind a bit! How is the quality of the TV?

HOUSTON: Oh, it's beautiful, Mike, it really is.

COLUMBIA: Oh, gee, that's great. Is the lighting half-way decent?

HOUSTON: Yes, indeed—they've got the flag up now and you can see the stars and stripes.

COLUMBIA: Beautiful, just beautiful!

*   *   *

HOUSTON: That's affirmative, Buzz. You're in our field of view now.

ALDRIN: All right. You do have to be rather careful to keep track of where your center of mass is. Sometimes it takes about two or three paces to make sure that you've got your feet underneath you. And about two to three, or maybe four, easy paces can bring you to a fairly smooth stop. Like a football player, you just have to put out to the side and cut a little bit. The so-called kangaroo hop—it does work, but it seems the forward ability is not quite as good as it is in the more conventional one foot after another. As far as saying what a safe pace might be—the one that I'm using now could get rather tiring after several hundred.

But this may be a function of the suit as well as lack of gravity forces.

HOUSTON: Tranquility Base, this is Houston. Could we get both of you in the camera for a minute, please?

TRANQUILITY BASE: Say again, Houston.

HOUSTON: Roger. We'd like to get both of you in the field of view of the camera for a minute. Neil and Buzz, the President of the United States is in his office now and would like to say a few words to you. Over.

TRANQUILITY BASE: That would be an honor.

[PRESIDENT/ASTRONAUT TALK HERE]

* * *

APOLLO CONTROL (12:06 A.M.): Neil is filling the bulk sample bag attached to a scale, which you see in the picture. Buzz is behind the LEM at the minus Z strut. That's the landing gear directly opposite the ladder. Neil has been on the surface about an hour and 10 minutes now.

EAGLE: We're now in the area of the minus Y strut.

APOLLO CONTROL: Buzz is making his way around the LEM, photographing it from various angles, looking at its condition on all sides. Neil is still occupied with the bulk sample. One hour 40 minutes expended on the PLSS's now.

ALDRIN: How's the bulk sample coming, Neil?

ARMSTRONG: The bulk sample is just being sealed.

* * *

ALDRIN: I think this right here is just as level.

ARMSTRONG: Okay.

APOLLO CONTROL: And they will be out of the camera's view while setting up these experiments.

ARMSTRONG: Look like salt and they have probably 2 per cent white minerals in 'em. White crystals. And the thing that I reported is—the . . . before. I don't believe I believe that anymore. I think it's small craters—they look like impact craters where shot—BB shot has hit the surface.

* * *

APOLLO CONTROL: Unofficial time off the surface at 111, 37, 32.

ALDRIN: Now start arching your back. That's good. Plenty of room. Arch your back a little. Roll right just a little bit. You're in good shape.

ARMSTRONG: Thank you.

ALDRIN: Now you're clear. You're rubbing up against me a little bit.

ARMSTRONG: Okay.

ALDRIN: That's right. A little to the left. Now move your foot and I'll get the hatch.

ALDRIN: Okay. The hatch is closed and latched. And verified secure.

ARMSTRONG: Okay.

Monday, July 21—from the lunar walk to reunion

HOUSTON (4 A.M.): Tranquility, we have a set of about 10 questions relating to observations you made or things you may have seen during the EVA. How do you feel? Over.

TRANQUILITY BASE: I think we can take a couple of them now.

HOUSTON: One of the implications here is the depth from which the bulk sample was collected. Did you manage to get down several inches, or near the surface?

TRANQUILITY BASE: We've got some down from as much as three inches in the area where I was looking at . . . variation with depth in the bulk sample really wasn't appreciable difference and I didn't run into any. . . . Later on, or at some other times in some other areas, why I dipped down a short distance—an inch or two—and couldn't go any further.

HOUSTON: Down as deep as three inches. Did not hit any hard bed. And no significant changes in composition to that depth.

TRANQUILITY BASE: We got two core tubes and a solar wind and about half of a big sample bag full of assorted rocks which I picked up hurriedly from around the area. I tried to get as many representative types as I could.

HOUSTON: On the two core tubes which you collected, how did the driving force required to collect these tubes compare? Was there any difference?

TRANQUILITY BASE: Not significantly. I could get down about the first two inches without much of a problem and then as I would pound it in about as hard as I could do it the second one took two hands on the hammer, and I was putting pretty good dents in the top of the extension rod. And it just wouldn't go much more than—I think the total depth might have been about 8 or 9 inches. But even there, it—for some reason—it didn't seem to want to stand up straight. So that I'd keep driving it in and it would dig some sort of a hole but it wouldn't just penetrate in a way that would support it and keep it from falling over. If that makes any sense at all. It didn't really to me.

TRANQUILITY BASE (Aldrin): And also I noticed when I took the bit off that the material was quite well packed, a good bit darker, and it—the way it adhered to the core tube gave me the distinct impression of being moist.

HOUSTON (4:20 A.M.): Tranquility Base, this is Houston. Two more verifications—can you, will you verify that the disc with messages was placed on the surface as planned, and, also, that the items listed in the flight plan —all of those listed there—were jettisoned.

TRANQUILITY BASE: All that's verified.

HOUSTON: Thank you, and I hope this will be a final goodnight.

APOLLO CONTROL (the NASA commentator, 10:31 A.M.): We've called the spacecraft Columbia from Mission Control here to wake up Mike Collins. The network is configured so that the LM [lunar module] crew—the Eagle crew—will not be disturbed.

HOUSTON: Hi, Mike, how's it going this morning?

COLUMBIA (Lieut. Col. Michael Collins in the command module, orbiting the moon): How goes it?

HOUSTON: Real fine. Real fine here.

APOLLO CONTROL (10:44 A.M.): The spacecraft Columbia has gone behind the moon on the 23d lunar revolution. Not since Adam has any human known such solitude as Mike Collins is experiencing during the 47 minutes of each lunar revolution when he's behind the moon with no one to talk to except his tape recorder aboard Columbia.

HOUSTON (11:12 A.M.): Tranquility Base, how's the resting standing up there? Did you get a chance to curl on the engine can?

TRANQUILITY BASE: Neil has rigged himself a very good hammock that . . . and he's been lying on the after engine cover and I curled up on the floor.

HOUSTON: I'll just read these notes on P22. Call P22 possible program alarm 526 range greater than 400 nautical miles and then use P22 as described on pings 20. Take option one in nouns zero six and use the no-update mode. The rendezvous radar will lock on at about 25 degrees elevation above the horizon. If 503 alarm occurs, designate fail. TA proceed and allow the rendezvous radar to search for the CSM. And place the range altitude monitor switch in altitude rate to prevent the tape meter from driving into the stops.

TRANQUILITY BASE: Roger. I think I have that.

HOUSTON (12:01 P.M.): Columbia, you're looking good to us.

COLUMBIA: Yes, sir. Keep it that way I got a crew status report from Columbia. I figure I got about five hours good sleep although you guys probably know better than I do.

HOUSTON (1:37 P.M.): You're cleared for take-off.

TRANQUILITY BASE: Roger. Understand. We're No. 1 on the runway.

APOLLO CONTROL (1:38 P.M.): We have confirmation on the ground that the ascent propulsion system propellant tanks have been pressurized.

TRANQUILITY BASE: Houston, we're not sure that we got No. 2 tank. Propellant is still showing a high pressure.

HOUSTON: We confirm that. Try it again.

TRANQUILITY BASE: We got No. 2 reading 3,050 and No. 1 is reading 3,000 and it drops down to 2,990. So I'm not sure it's really indicative that it didn't go.

HOUSTON: We copy and we agree.

TRANQUILITY BASE: Okay. I assume we're go for lift-off and will proceed with the ascent feeds.

HOUSTON (1:44 P.M.): A little less than 10 minutes here. Everything looks good.

HOUSTON (1:50 P.M.): Eagle [the lunar module in space], you're looking good to us. We'll continue to monitor now at 3 minutes 12 seconds away from ignition as crew of Eagle goes through their pre-launch checklist. Guidance reports both navigation systems on Eagle are looking good.

TRANQUILITY BASE: Nine, eight, seven, six, five, first stage engine on ascent. Proceed. Beautiful, 26, 36 feet per second up. Little pitch over, very smooth, very quiet ride. There's that one crater down there.

HOUSTON (1:54 P.M.): A thousand feet high, 80 feet per second vertical rise.

HOUSTON: Twenty-six hundred feet altitude. Eagle, Houston. One minute and you're looking good. A hundred thirty feet vertical rise rate.

EAGLE: A little bit of slow wallowing back and forth. Not very much thruster activity.

HOUSTON: Mighty fine.

HOUSTON: Eagle, you're go at three minutes. Everything's looking good.

EAGLE: Right. This is . . . This is H dot max now. Right down U.S. 1.

APOLLO CONTROL: Height's now approaching 32,000 feet.

HOUSTON: Eagle, four minutes. You're going right down the track. Everything is great.

EAGLE: Horizontal velocity approaching 2,500 feet per second. We've got Sabine now to our right now.

APOLLO CONTROL: Some 120 miles to go until insertion.

HOUSTON: (2 P.M.): Eagle, Houston. You're still looking mighty fine.

EAGLE: Eagle is back in orbit, having left Tranquility Base, and leaving behind a replica from our Apollo 11 patch with an olive branch.

HOUSTON: Roger. We copy. The whole world is proud of you.

APOLLO CONTROL (3:26 P.M.): Less than a minute away from acquisition of the spacecraft Columbia coming around on the near side of the moon on the 26th revolution. Some three minutes, 11 seconds away from Eagle's appearance on the lunar frontside. We have AOS [acquisition of signal] of the spacecraft Columbia.

APOLLO CONTROL: Range between Eagle and Columbia now showing 67.5 nautical miles. Range rate—closure rate—121 feet per second.

APOLLO CONTROL (4:08 P.M.): The Black Team of Apollo Mission control are more or less in an advisory capacity during this rendezvous sequence. They're actively computing maneuver times but in the final analysis it's on-board computations by the crew of Columbia and Eagle which really bring about the rendezvous.

EAGLE (4:31 P.M.): Mike, if you want our target delta V, I'll give it to you.

COLUMBIA: Ready to copy.

EAGLE: 127 03 3082 plus 22.7 plus 1.7 minus 10.6.

COLUMBIA: 127 03 3082 plus 22.7 plus 1.7 minus 10.6. Thank you.

APOLLO CONTROL (5:23 P.M.): Less than a minute now away from acquisition of the spacecraft Columbia. Hopefully flying within a few feet of it will be Eagle. Docking should take place about 10 minutes from now according to the flight plan. However, this is a crew-option matter.

EAGLE (5:27 P.M.): Okay, Mike. I'll try to get positioned here. Then you got it.

COLUMBIA: How did the roll attitude look? I'll stop. As a matter of fact, I could stop right here, if you like that.

EAGLE (5:30 P.M.): I'm not going to do a thing, Mike. I'm just letting her hold in attitude. . . .

COLUMBIA: Okay.

EAGLE (5:36 P.M.): We're all yours, Columbia.

COLUMBIA: Okay.

UNIDENTIFIED VOICES: I'm pumping out cabin pressure.

That was a funny one. You know, I didn't feel . . . and I thought things were pretty steady. I wanted to retract . . . when all hell broke loose. Were you guys—did it appear to you to be that you were jerking around quite a bit during your retract cycle?

Yeah. It seemed to happen at the time I put the Plux-X thruster up. It apparently—it wasn't finished because somehow or other I . . . to get off in attitude. . . .

Yeah, I was sure busy there for a couple of seconds.

Could you hear me all right? I got a horrible squeal.

Yeah, I agree with that. But we read you okay.

APOLLO 11 (the three-man crew, after link-up) (5:38 P.M.): Houston, Apollo 11. Over.

HOUSTON: Apollo 11, Houston, Go.

APOLLO CONTROL (6:36 P.M.): We've had loss of signal as Apollo 11 went around the back side of the moon. Armstrong and Aldrin should be involved in preparing for transferring back to the command module with Mike Collins and they'll be cleaning up equipment and vacuuming off any particles of dust that remain before transferring to the command module.

HOUSTON (7:20 P.M.): Hello Eagle, Houston. Do you read?

COLUMBIA: Read you loud and clear. We're all three back inside. The hatch is installed. We're ready to . . . Everything's going well.

U.S. and Soviet Manned Space Flights

SPACECRAFT	LAUNCHING DATE	ASTRONAUT	REVOLUTIONS	FLIGHT TIME	FLIGHT HIGHLIGHTS
U.S.S.R. Vostok 1	Apr. 12, 1961	Yuri A. Gagarin	1	1 hr. 48 mins.	First manned flight.
U.S. Mercury-Redstone 3	May 5, 1961	Alan B. Shepard, Jr.	Sub-orbital	0 hrs. 15 mins.	First American in space.
U.S. Mercury-Redstone 4	July 21, 1961	Virgil I. Grissom	Sub-orbital	0 hrs. 16 mins.	Capsule sank.
U.S.S.R. Vostok 2	Aug. 6, 1961	Gherman S. Titov	16	25 hrs. 18 mins.	More than 24 hours in space.
U.S. Mercury-Atlas 6	Feb. 20, 1962	John H. Glenn, Jr.	3	4 hrs. 55 mins.	First American in orbit.
U.S. Mercury-Atlas 7	May 24, 1962	M. Scott Carpenter	3	4 hrs. 56 mins.	Landed 250 miles from target.
U.S.S.R. Vostok 3	Aug. 11, 1962	Andrian G. Nikolayev	60	94 hrs. 22 mins.	First group flight. (Vostok 3 and 4)
U.S.S.R. Vostok 4	Aug. 12, 1962	Pavel R. Popovich	45	70 hrs. 57 mins.	Came within 3.1 miles of Vostok 3 on first orbit.
U.S. Mercury-Atlas 8	Oct. 3, 1962	Walter M. Schirra, Jr.	6	9 hrs. 13 mins.	Landed 5 miles from target.

U.S. and Soviet Manned Space Flights—(Continued)

SPACECRAFT	LAUNCHING DATE	ASTRONAUT	REVOLUTIONS	FLIGHT TIME	FLIGHT HIGHLIGHTS
U.S. Mercury-Atlas 9	May 15, 1963	L. Gordon Cooper, Jr.	22	34 hrs. 20 mins.	First long flight by an American.
U.S.S.R. Vostok 5	June 14, 1963	Valery F. Bykovsky	76	119 hrs. 6 mins.	Second group flight. (Vostok 5 and 6)
U.S.S.R. Vostok 6	June 16, 1963	Valentina V. Tereshkova	45	70 hrs. 50 mins.	Passed within 3 miles of Vostok 5; first woman in space.
U.S.S.R. Voskhod 1	Oct. 12, 1964	Vladimir M. Komarov Konstantin P. Feoktistov Dr. Boris G. Yegorov	15	24 hrs. 17 mins.	First 3-man craft.
U.S.S.R. Voskhod 2	Mar. 18, 1965	Aleksei A. Leonov Pavel I. Belyayev	16	26 hrs. 2 mins.	First man outside spacecraft in 10-minute "walk" (Leonov).
U.S. Gemini 3	Mar. 23, 1965	Virgil I. Grissom John W. Young	3	4 hrs. 53 mins.	First manned orbital maneuvers.

U.S. Gemini 4	June 3, 1965	James A. McDivitt / Edward H. White, 2nd	62	97 hrs. 48 mins.	21-minute "space walk" (White).
U.S. Gemini 5	Aug. 21, 1965	L. Gordon Cooper, Jr. / Charles Conrad, Jr.	120	190 hrs. 56 mins.	First extended manned flight.
U.S. Gemini 7	Dec. 4, 1965	Frank Borman / James A. Lovell, Jr.	206	330 hrs. 35 mins.	Longest space flight.
U.S. Gemini 6-A	Dec. 15, 1965	Walter M. Schirra, Jr. / Thomas P. Stafford	16	25 hrs. 52 mins.	Rendezvous within 1 foot of Gemini 7.
U.S. Gemini 8	Mar. 16, 1966	Neil A. Armstrong / David R. Scott	6.5	10 hrs. 42 mins.	First docking to Agena target; mission cut short.
U.S. Gemini 9-A	June 3, 1966	Thomas P. Stafford / Eugene A. Cernan	44	72 hrs. 21 mins.	Rendezvous, extra-vehicular activity, precision landing.
U.S. Gemini 10	July 18, 1966	John W. Young / Michael Collins	43	70 hrs. 47 mins.	Rendezvous with 2 targets; Agena package retrieved.

U.S. and Soviet Manned Space Flights—(Continued)

SPACECRAFT	LAUNCHING DATE	ASTRONAUT	REVOLUTIONS	FLIGHT TIME	FLIGHT HIGHLIGHTS
U.S. Gemini 11	Sept. 12, 1966	Charles Conrad, Jr. Richard F. Gordon, Jr.	44	71 hrs. 17 mins.	Rendezvous and docking.
U.S. Gemini 12	Nov. 11, 1966	James A. Lovell, Jr. Edwin E. Aldrin, Jr.	59	94 hrs. 33 mins.	3 successful extra-vehicular trips.
U.S.S.R. Soyuz 1	Apr. 23, 1967	Vladimir M. Komarov	17	26 hrs. 40 mins.	Heaviest manned craft; crashed, killing Komarov.
U.S. Apollo 7	Oct. 11, 1968	Walter M. Schirra, Jr. Donn F. Eisele R. Walter Cunningham	163	260 hrs. 9 mins.	First manned flight of Apollo spacecraft.
U.S.S.R. Soyuz 3	Oct. 26, 1968	Georgi T. Beragovoi	60	94 hrs. 51 mins.	Rendezvous with unmanned Soyuz 2.
U.S. Apollo 8	Dec. 21, 1968	Frank Borman James A. Lovell, Jr. William A. Anders	Moon orbital (10 revolutions)	147 hrs.	First manned voyage around moon.

U.S.S.R. Soyuz 4	Jan. 15, 1969	Vladimir A. Shatalov	45	71 hrs. 14 mins.	Rendezvous with Soyuz 5.
U.S.S.R. Soyuz 5	Jan. 15, 1969	Boris V. Volynov Aleksei S. Yeliseyev Yevgeny V. Khrunov	46	72 hrs. 46 mins.	Rendezvous with Soyuz 4; Yeliseyev and Khrunov transfer to Soyuz 4.
U.S. Apollo 9	Mar. 3, 1969	James A. McDivitt David R. Scott Russell L. Schweickart	151	241 hrs. 1 min.	Docking with Lunar Module.
U.S. Apollo 10	May 18, 1969	Thomas R. Stafford Eugene A. Cernan John W. Young	Moon orbital (31 revolutions)	192 hrs. 3 mins.	Descent to within 9 miles of moon.
U.S. Apollo 11	July 16, 1969	Neil A. Armstrong Edwin E. Aldrin, Jr. Michael Collins	Moon orbit for command module (31 revolutions) Time spent on moon 21 hrs. 38 mins.	195 hrs. 18 mins.	Armstrong and Aldrin, in lunar module, land and walk on moon.

Major Apollo/Saturn V Contractors
As Compiled by NASA

Contractor	Item
Bellcomm Washington, D.C.	Apollo Systems Engineering
The Boeing Co. Washington, D.C.	Technical Integration and Evaluation
General Electric Apollo Support Dept. Daytona Beach, Fla.	Apollo Checkout, and Quality and Reliability
North American Rockwell Corp. Space Div. Downey, Calif.	Command and Service Modules
Grumman Aircraft Engineering Corp. Bethpage, N.Y.	Lunar Module
Massachusetts Institute of Technology Cambridge, Mass.	Guidance & Navigation (Technical Management)
General Motors Corp. AC Electronics Div. Milwaukee, Wis.	Guidance & Navigation (Manufacturing)
TRW, Inc. Systems Group Redondo Beach, Calif.	Trajectory Analysis LM Descent Engine LM Abort Guidance System
Avco Corp. Space Systems Div. Lowell, Mass.	Heat Shield Ablative Material
North American Rockwell Corp. Rocketdyne Div. Canoga Park, Calif.	J-2 Engines, F-1 Engines
The Boeing Co. New Orleans, La.	First Stage (SIC) of Saturn V Launch Vehicles, Saturn V Systems Engineering and Integration, Ground Support Equipment
North American Rockwell Corp. Space Div. Seal Beach, Calif.	Development and Production of Saturn V Second Stage (S-II)
McDonnell Douglas Astronautics Co. Huntington Beach, Calif.	Development and Production of Saturn V Third Stage (S-IVB)

317

Contractor	Item
International Business Machines Federal Systems Div. Huntsville, Ala.	Instrument Unit
Bendix Corp. Navigation and Control Div. Teterboro, N.J.	Guidance Components for Instrument Unit (Including ST-124M Stabilized Platform)
Federal Electric Corp. Paramus, N.J.	Communications and Instrumentation Support, KSC
Bendix Field Engineering Corp. Owings Mills, Md.	Launch Operations/Complex Support, KSC
Catalytic-Dow Titusville, Fla.	Facilities Engineering and Modifications, KSC
Hamilton Standard Div. United Aircraft Corp. Windsor Locks, Conn.	Portable Life Support System; LM ECS
ILC Industries Dover, Del.	Space Suits
Radio Corp. of America Van Nuys, Calif.	110A Computer—Saturn Checkout
Sanders Associates Nashua, N.H.	Operational Display Systems Saturn
Brown Engineering Huntsville, Ala.	Discrete Controls
Reynolds, Smith and Hill Jacksonville, Fla.	Engineering Design of Mobile Launchers
Ingalls Iron Works Birmingham, Ala.	Mobile Launchers (ML) (structural work)
Smith/Ernst (Joint Venture) Tampa, Fla. Washington, D.C.	Electrical Mechanical Portion of MLs
Power Shovel, Inc. Marion, Ohio	Transporter
Hayes International Birmingham, Ala.	Mobile Launcher Service Arms
Bendix Aerospace Systems Ann Arbor, Mich.	Early Apollo Scientific Experiments Package (EASEP)
Aerojet-Gen. Corp. El Monte, Calif.	Service Propulsion System Engine

. . . AND . . .

Thousands of subcontractors across the country, employing tens of thousands of people.

Bibliography

This is not a complete listing of the research materials that went into this book. That would be impossible, since so much of it came from personal interview. But the following books were especially helpful and would be useful to anyone wanting to pursue in detail a study of the Apollo Project.

Alexander, Tom W., *Project Apollo*. New York: Harper & Row, 1964.

Branley, Franklyn M., *Exploration of the Moon*. Garden City, N. Y.: Doubleday, 1964.

Clarke, Arthur C., *Man and Space*. New York: Life Science Library, 1966.

——————————, *The Promise of Space*. New York: Harper & Row, 1968.

Diamond, Edwin, *The Rise and Fall of the Space Age*. Garden City, N. Y.: Doubleday, 1964.

Grissom, Virgil I., *Gemini: A Personal Account of Man's Venture into Space*. New York: Macmillan, 1968.

Holmes, Jay, *America on the Moon*. Philadelphia: Lippincott, 1962.

Hunter, Maxwell W., *Thrust into Space*. New York: Holt, Rinehart & Winston, 1965.

Kennan, Erlend A. and Harvey, Edmund H., Jr., *Mission to the Moon*. New York: Morrow, 1969.

Lewis, Richard S., *Appointment on the Moon*. New York: Viking, 1968.

Ley, Willy, *Rockets, Missiles and Men in Space*. New York: Viking, 1967.

McGraw-Hill Encyclopedia of Space, New York: McGraw-Hill, 1968.

Rosholt, Robert C., *An Administrative History of NASA, 1958-1963*. Washington: National Aeronautics and Space Administration, 1966.

Shelton, William, *Soviet Space Exploration: The First Decade*. New York: Washington Square Press, 1968.

Sullivan, Walter (editor), *America's Race for the Moon.* New York: Random House, 1962.

Swenson, Lloyd S., Jr., et al, *This New Ocean: A History of Project Mercury.* Washington: National Aeronautics and Space Administration, 1966.

Taylor, L. B., Jr., *Liftoff: The Story of America's Spaceport.* New York: Dutton, 1968.

Von Braun, Wernher and Ordway, Frederick I., *History of Rocketry and Space Travel.* New York: Crowell, 1967.

Whipple, Fred L., *Earth, Moon and Planets.* Cambridge, Mass.: Harvard University Press, 1963.

Wright, Hamilton and Helen, and Rapport, Samuel (editors), *To the Moon.* New York: Meredith, 1968.

Index

A NOTE ON THIS BOOK

The text of this book is set in 9 point Times Roman. This typeface was commissioned by the *Times* of London in 1931 and the design was supervised by Stanley Morison, the typographic adviser to the Monotype Corporation, Ltd., for production on the linotype in 1932. It is a highly legible and versatile masculine typeface and is the most popular text type design in America today. The display type is set in Century Bold, a variant of the Century typeface family. This face derives from Century Roman which was cut by L. B. Benton in 1895 under the guidance of Theodore L. DeVinne.

The compositors for the text of *We Reach the Moon* are the Monotype Composition Company, Baltimore, Maryland, and the Central Typesetting and Electrotyping Company, Chicago, Illinois.

The cover engraving and printing are by The Regensteiner Press, Chicago, Illinois.

The color insert engravings are from Colormatic Litho, Incorporated, Chicago, Illinois.

Text printing, color insert printing and binding are by the W. F. Hall Printing Company, Chicago, Illinois.

Paper was supplied by Boise Cascade Papers and Bergstrom Paper Company.